Defending Freedom

Defending Freedom

How We Can Win the Fight for an Open Society

Ralf Fücks

Translated by Nick Somers

polity

First published in German as *Freiheit Verteidigen* © Carl Hanser Verlag GmbH. & Co. KG 2017

This English edition © Polity Press, 2019

Polity Press
65 Bridge Street
Cambridge CB2 1UR, UK

Polity Press
101 Station Landing
Suite 300
Medford, MA 02155, USA

All rights reserved. Except for the quotation of short passages for the purpose of criticism and review, no part of this publication may be reproduced, stored in a retrieval system or transmitted, in any form or by any means, electronic, mechanical, photocopying, recording or otherwise, without the prior permission of the publisher.

ISBN-13: 978-1-5095-3623-8
ISBN-13: 978-1-5095-3624-5 (pb)

A catalogue record for this book is available from the British Library.

Library of Congress Cataloging-in-Publication Data
Names: Fuecks, Ralf, author.
Title: Defending freedom : how we can win the fight for an open society / Ralf Fuecks.
Other titles: Freiheit verteidigen. English
Description: English edition. | Cambridge ; Medford, MA : Polity Press, [2019] | Includes bibliographical references and index. |
Identifiers: LCCN 2019011787 (print) | LCCN 2019017016 (ebook) | ISBN 9781509536252 (Epub) | ISBN 9781509536238 (hardback) | ISBN 9781509536245 (pbk.)
Subjects: LCSH: Liberty. | Liberalism. | Democracy. | European Union.
Classification: LCC JC585 (ebook) | LCC JC585 .F8313 2019 (print) | DDC 320.01/1--dc23
LC record available at https://lccn.loc.gov/2019011787

Typeset in 11 on 13 Sabon by Fakenham Prepress Solutions, Fakenham, Norfolk, NR21 8NL
Printed and bound in Great Britain by T J International Ltd

The publisher has used its best endeavours to ensure that the URLs for external websites referred to in this book are correct and active at the time of going to press. However, the publisher has no responsibility for the websites and can make no guarantee that a site will remain live or that the content is or will remain appropriate.

"Children's Anthem", originally published in German in 1950 as "Kinderhymne". Copyright © 1950, 1964, 1976 by Bertolt-Brecht-Erben / Suhrkamp Verlag, from BERTOLT BRECHT POEMS 1913-1956 by Bertolt Brecht, edited by John Willet and Ralph Manheim. Used by permission of Liveright Publishing Corporation.

Every effort has been made to trace all copyright holders, but if any have been overlooked the publisher will be pleased to include any necessary credits in any subsequent reprint or edition.

For further information on Polity, visit our website: politybooks.com

Contents

1 In Place of an Introduction: The Lie of the Land 1
2 Modern and Anti-Modern 17
3 The Long View of Democracy 24
4 The Left and Democracy 48
5 The Rise of the Anti-Liberals 59
6 The Migration Battlefield 68
7 Dealing with Islam 78
8 No Empathy for Freedom: The Germans and Ukraine 95
9 The Russian Complex 100
10 Modernity and Its Discontents 110
11 Ecology and Freedom 129
12 Civilizing Capitalism 150
13 Shaping Globalization 165
14 How We Can Relaunch the European Union 174
15 What is at Stake 179

Notes 197

The greatest threat always comes from within the West itself, from a West which denies its values, which it has done often enough.

Heinrich August Winkler

If we want to stay human, there is then only one path, the path towards an open society. We must step into the unknown, into the uncertain and into the precarious; we must use the reason that lies at our disposal to secure both certainty and freedom.

Karl Popper

1

In Place of an Introduction: The Lie of the Land

A spectre is haunting the western world: a revolt against the open society. It has many faces and forms. In the United States, a rank outsider and his aggressive rhetoric made it to the White House. His election campaign was based on resentment – of migrants and free trade, feminists and Muslims, 'Washington' and the East Coast and Californian left-wing liberal elites. Donald Trump's election victory turned everything on its head. An egocentric New York billionaire became the voice of an anxious and enraged America, a hero of the white working class and provincial America. He prevailed over a large part of the Republican establishment, over the outcry from cultural elites and over all conventional wisdom that elections are won in the centre ground. Trump polarized the country, regardless of the wounds he opened up. He did not require a perfectly orchestrated election campaign because he tuned into the emotions of millions of Americans who could not identify with the rainbow coalition of left-wing liberal elites and sexual and ethnic minorities. What remains is the bitter recognition that, in spite or even because of his demonizing of migrants, his misogynistic outbursts, his reckless foreign policy and his vanity and narcissism, he was able to advance to become president of what is still the most powerful country in the world: a man who cares as little about America as the bastion of liberty and free trade as he does about the West as a community of shared values. Since his election, Trump has shattered all illusions that the newcomer in the White House would be reined in by the professional politicians in the administration and in Congress. The first year of his presidency saw him rampage increasingly against America's liberal tradition, transatlantic relations and a multilateral world order created to a large extent by the United States after the Second World War.

We Europeans have no cause to point the finger at America. Europe has also long felt the reverberations of the crisis in liberal democracy. From Scandinavia to Italy, from France to Poland, anti-liberal parties have risen to prominence. They have shifted public discourse to the right and pushed the established parties towards more extreme positions. Nationalist, xenophobic and anti-European forces are now part of the governments in Poland, Hungary, Austria and, most recently, Italy. Even in Germany, hitherto an anchor of stability in Europe, the dam has burst to let in extreme right-wing tendencies. An underlying feeling of anxiety, fear of the future and diffuse anger is spreading. It forms a springboard for radical movements and political demagogues and for the call for a strong state that secures its borders and screens society from the ravages of globalization.

However different they may be as individuals, Donald Trump, Marine Le Pen, Viktor Orbán, Jarosław Kaczyński and their like-minded associates have some fundamental things in common. They claim to represent the voice of the people against an elite that is out of touch with reality; they appeal to strong feelings and passions – patriotism, identity, fear, envy – and they make a clear distinction between 'us' and 'them', friend and foe. These are all main characteristics of populist politics. It is not only their political ideas that are dangerous. They are dangerous above all because they attack the institutional structure on which a democratic republic is based: the independence of the judiciary; freedom of the media; the impartiality of the public administration; and the protection of minorities. They conceive politics as a latent civil war in which the victors grab all of the power. The term 'illiberal democracy' is being bandied about to describe a form of government that is ostensibly democratic but in reality authoritarian, one in which the majority party can attack the foundations of the liberal rule of law – separation of powers, political pluralism, minority rights and cultural diversity.

In terms of ideology, it is a mixture of nationalistic, conservative, populist (*völkisch*) and socialist elements. The anti-liberals invoke national sovereignty and call for direct democracy as a means of attacking the political establishment. The new right borrows unashamedly from the traditional left – it styles itself as the protector of 'ordinary people', promises to defend domestic workers from the onslaught of globalization and calls for the priority of politics over markets. And it joins forces with a nationalistic left in its resentment of the United States and its indictment of 'international finance capital' and its call for the recovery of national sovereignty. The old fronts are becoming blurred when it comes to mobilization against the Transatlantic Trade and Investment Partnership (TTIP), protection of rural agricultural workers, general suspicion of NATO or sympathy for the politics of Vladimir Putin. It is no coincidence that most of the right- and left-wing populist parties in Europe are enthusiastic supporters of the autocratic regime in Moscow.

In early 2016, the right-wing nationalist candidate narrowly failed in his bid to be elected president of Austria. This was the prelude to the thunderbolt that shook all of Europe on 23 June 2016. On that noteworthy day, 51.9 per cent of the British public voted in favour of leaving the European Union (EU). As in Austria, a clear majority of elderly persons, workers and the rural population revolted against the political credo of the liberal elites.

The Brexit alliance embraced the entire political spectrum, from the far right to the far left. It was driven by a mixture of British nostalgia, fear of uncontrolled mass migration and a call for recovery of national sovereignty, coupled with a grotesquely exaggerated idea of the power of the bureaucrats in Brussels. Added to this was the illusion that the United Kingdom could ride the waves of globalization better on its own. As usual in such cases, domestic motives were mixed up with the condemnation of the EU. In the former industrial centres of England in particular, which had been among the losers in Margaret Thatcher's neo-liberal revolution, the referendum offered a welcome opportunity to give vent to the stored-up bitterness towards 'those in power'. Direct democracy, once a leftist ideal, has now become a weapon used by the disadvantaged and those who feel insecure or ignored in their fight against the cosmopolitan elites. A new cultural struggle has broken out in the centres of the western world. Issues long thought to have been settled – multiculturalism, religious pluralism, the end of patriarchy, sexual diversity, the embeddedness of national politics in multilateral institutions – are once again being called into question.

The West in this context is more than a geographical designation. As a political category, it stands for the modern liberalism project. It covers an area shaped by the Reformation, Enlightenment and the notion of human rights. The Bill of Rights (1689), the Declaration of Independence (1776) and the Declaration of the

Rights of Man and of the Citizen (1789) are founding documents of modern democracy. Freedom and equality are their guiding principles, the separation of powers, the rule of law, separation of state and religion and documented civil rights are their constituent elements. The era in which global politics and trade were dominated by the West is coming inexorably to an end. At stake now is the survival of the West as a democratic community of shared values – and with it the liberal universalism manifested in the Charter of the United Nations and the Universal Declaration of Human Rights.

Anti-liberal cross-party coalition

Across Europe, there are striking points of contact between right-wing nationalists and left-wing sovereigntists. They see the European Union as a Trojan horse of 'neo-liberal globalization'. They claim to be European but on the basis of a 'Europe of nations' and national self-determination. Antipathy to NATO as a representative of American hegemony goes hand in hand with sympathy for Vladimir Putin and his anti-western politics. Moscow today is in fact the headquarters of a pan-European network of antiliberal parties, associations and media. They reject the idea of universal values and regard the notion that human rights and democracy should be global as a liberal pretension.

After the failed interventions in Afghanistan and Iraq, public opinion is averse to risks and seeks to avoid conflicts. By contrast, the Russian leadership relies on the calculated use of force as a political instrument. The large-scale offensive military manoeuvres and the provocative actions of Russian bombers and the threatened use of tactical nuclear weapons are designed to spread fear of a major war and to intimidate Europeans: a conflict with Russia is the last thing they want! At the same time, the Kremlin is expanding its propaganda networks. State-owned media like *Russia Today* and *Sputnik* are instruments of ideological warfare abroad, and a whole army of trolls and hackers are active partisans in the information war. Added to that is a wide-ranging network of foundations, institutes and think tanks in Europe, some controlled directly by Moscow, others in cooperation with Putin's friends in Russia. Former top European politicians are lured into taking lucrative positions on supervisory boards and in foundations, not least a former German federal chancellor, who does not tire of speaking up for Putin. State-owned Russian companies acquire key positions in the European energy industry; a legion of financial institutions, law firms, real estate companies and PR agencies are involved in business done with Russian money. And finally there are the close connections between the Kremlin and both right-wing and left-wing parties in Europe.

For all the colourful diversity of the cross-party front of anti-western groups, what they have in common is their opposition to liberalism and globalization. They counter the notion of liberal universalism with the concept of a multipolar world order, characterized by large areas, each with their own traditions and norms: diversity on a global scale; and homogeneity at home. They fear the Islamization of Europe but grant Islam its own sphere of influence, in which the West should not interfere. In this ethno-pluralistic world, Russia forms a civilization in its own right – an uncontrollable counterpoint to the ideological and political domination of the West. Left- and right-wing actors also share an admiration for China. Surely the Middle Kingdom has the right to assume its historical role as a world power again? And surely China's massive economic and social rise justifies its authoritarian form of government? In the eyes of many western observers as well, political stability and economic strength count more than democratic rights and freedoms. Sympathy for the Chinese approach even extends far into the environmental sector. Is a 'benevolent dictatorship' not in a better position to take the necessary measures regarding production and consumption than western democracies with their fixation on the next elections?

The offshoots of this anti-liberal counter-revolution can be felt right across the party spectrum. They are nourished by fear of globalization and of losing status, the sense that individuals no longer have control over their lives, or indeed anything else, a concoction of xenophobia mixed with toxic social envy, growing mistrust of the political and business elites and a basic feeling of being overwhelmed by the speed of cultural, technical and social change. The fear of being swamped by immigrants of a different colour and religion is just the tip of a huge iceberg of insecurity. Globalization and the digital revolution have led to a growing polarization of our society into winners and losers. Those who have acquired a good education, speak several languages, have international contacts and are technically literate are more likely to see open markets, migration and cultural diversity as an opportunity. The rest are more likely see them as a threat.

It is against this background that populist movements have been growing on both left and right. They claim to be the true voice of the people against the 'regime parties' (*Systemparteien*) and 'regime press' (*Systempresse*). It is no coincidence that these battle cries from the 1920s and 1930s are now being revived. The opposition of both left- and right-wing parties to the system is also nothing new. The idea of a revolutionary cross-party front was already popular in the nationalist communist and conservative circles in the Weimar Republic, as was the demand for a Berlin–Moscow axis as a counter-balance to London and Washington. The confrontation with anti-liberal movements today is not taking place on virgin soil. The revolt against modern liberalism and anti-western attitudes is deeply rooted in European thinking.[1]

But, of course, history does not repeat itself, and caution is required when drawing historical parallels. Democracy in Germany today is far more stable than it was at any time during the Weimar Republic. But no nation is immune to the return of anti-democratic movements. In many European countries, they already attract 20–30 per cent of the electorate. The common denominator is contempt for liberal democracy, return to the nationalist fold, defence of an imagined cultural homogeneity and evocation of family, nation and state as the bastions against the threat from without.

Society, seen as cold, impersonal and alienating, is contrasted with a *community* longing for a sense of belonging; the abstractness of the market with the ideal of a local economy based on personal relations; the remoteness of representative democracy with the directness of a plebiscite; the excessive demands of a multicultural society with the desire for homogeneity; liberal universalism with the idea of a plural world order in which every cultural group lives according to its own values.

These regressive tendencies are by no means confined to those who have not been able to reap the benefits of modernization or have been left behind. The new feature of the anti-liberal revolt is that it has spread both vertically and horizontally. It also includes bourgeois circles and parts of the left. The 'enraged citizen' (*Wutbürger*) is generally well educated, pursues a recognized profession and is not among the poorest. He still has a comfortable life, but he feels the ground shaking under his feet. Economically, he experiences growing pressure and competition. Culturally, he feels threatened by the patriarchal crisis, the loss of male security, the blatancy of gays and lesbians and immigration from Muslim countries. He has the impression that there is money available for everything, just not for him and the things that concern him. He feels abandoned and bossed around by 'those in power', and he is unhappy in general about the way things are going.

The new radical chic

The resurgent longing for grand gestures and great deeds, for the scent of danger and community, is not confined to the New Right press, which reverberates with the sound of resistance, revolt and revolution. The dalliance with radical chic can also be found in the feature sections of the main media, in the fundamental tone of opposition to the system in the world of art and culture, the star cult surrounding theoretical acrobatics of neo-communist contortionists like Slavoj Žižek, or the never-ending fascination that appears to emanate from the revolutionary habitus of the radical left.[2]

When everyday life becomes boring and the middle-class male has finally been pacified, a growing longing for heroism, for breaking taboos and for flirtations with danger emerges from underneath the politically correct surface. Rammstein, the most successful German band of the last twenty years, recognized this emerging zeitgeist early on and gave it a form. Now it has crossed over from the world of culture into the world of politics, and what do we have? Now it is the New Right that has adopted the habitus of subversion, non-compliance and resistance, making the by now bourgeois left look pale in comparison. The politics of anger has shifted to the right. The politician and trained philosopher Marc Jongen, a spokesman for Alternative für Deutschland (AfD), talks of the repoliticization of pride and anger by the New Right and calls for militant opposition to radical Islam. Wherever 'German culture' is threatened, resistance becomes a duty. The sound is familiar, the content disconcerting.

At a time when consumer capitalism is being restrained by the welfare state, talk of the *Verhausschweinung des Menschen* (literally, turning people into domesticated pigs) is being heard again, somewhat predictably, in some circles. The term was coined by Konrad Lorenz, Germany's favourite behavioural scientist. It refers to the domestication of mankind and the neutralization of the mechanisms of natural selection (survival of the fittest). Lorenz saw this as a degenerative process. The social Darwinist element in this theory is evident. It has a long tradition. 'In peace,' said Georg Wilhelm Friedrich Hegel, 'the bounds of civil life are extended ... and in the long run, people become stuck in their ways.' The only remedy is the purifying storm of war, the antidote to 'political nullity and boredom'. This was the mood in which millions of Germans perceived the start of the First World War as a form of liberation from the constraints of bourgeois life. The anti-bourgeois dalliance with danger and violence awakens spirits that outgrow their creators.

The *emasculation of politics* is a classic topos of cultural criticism from both the left and the right.[3] They complain about the loss of the fighting, protective and existential elements in politics. Pragmatism, anti-war sentiment and willingness to compromise are seen as symptoms of weakness. Karl Popper, one of the twentieth century's most outstanding liberal philosophers, said on this subject, 'The tribal ideal of the Heroic Man . . . is an attack on the idea of civil life itself; this is denounced as shallow and materialistic. . . . live dangerously! is its imperative.'[4]

The attraction which communism and fascism has for intellectuals and artists is due not least to the aura of violence of revolutionary movements, their heroic gestures and righteous solemnity. Politics that does not go all the way, parliamentary compromise and pragmatic 'piecework technology' (Popper) are less attractive. It is no coincidence that the New Right refers to Carl Schmitt again: politics must be considered from the point of view of a state of emergency; in essence it is a struggle between friend and foe. The equivalent on the left is the romance of violence, symbolized even today by Che Guevara, whose death in the guerrilla war merely heightened his cult status. The fascinating feature of the icons of left-wing radicalism is their uncompromising habitus in which *humanism* combines with *terror* – following the lines in the famous essay by Maurice Merleau-Ponty, who celebrated the violence of the oppressed as an act of revolt through which they regain their dignity as human beings.

Fear eats the soul

While other continents are moving towards a better future full of self-confidence and dynamism, Europeans have become disenchanted by modernity. Europe today is the continent most fearful of the future. Nowhere else in the world is there such a widespread belief that the golden years lie in the past. The fear of globalization and free trade, of the digital revolution and gene technology, of mass immigration and Islamization, of climate change and old-age poverty outweighs optimism for the future. The growth dynamic is weak, and in many countries the level of youth unemployment is alarming. The optimism of 1989/90 has been transformed into mutual discontentment and national egoism. This plays into the hands of the defenders of sovereignty on both the left and the right, who demand a retreat into the fortress of the nation-state. The future they promise the insecure masses is to be found in a return to an idealized past.

Nothing illustrates Europe's self-doubt more clearly than the fearful attitude towards the influx of refugees from war and poverty. A large part of the population (and the political elites) apparently no longer believe in the persuasiveness of our way of life and the integrative dynamism of democracy and the market economy. Even in the United States, a society of immigrants, there is a strident call for isolation from the havenots to the south. What looks like a patriotic display of strength is in reality a sign of status panic.

Eurasian community versus Atlantic Europe

On top of the internal crisis of western democracies is the challenge from without. Gone is the era when Francis Fukuyama announced the 'end of history' and the whole world appeared to be on the way towards democracy and the market economy. It makes no sense any more to call countries like Russia, China or Iran 'societies in transition' converging with the western model. They are authoritarian systems in their own right that with growing self-confidence offer an alternative to the West. Russia with Putin at the helm is navigating a decidedly anti-western course. The Kremlin is not only extending its geopolitical sphere of influence but is also waging an ideological campaign against liberalism. Russia styles itself as the last bastion of Christian values, as opposed to a decadent 'Gayropa' colonized by America. Greater Russia ideologists like Alexander Dugin[5] are re-warming the old conflict dichotomy of culture and civilization. They contrast tragic heroism, a willingness to make sacrifices, ethnic communities and a nation of culture with a degenerate consumer society that indulges in corrosive hedonism and gives itself up to worship of the Golden Calf, the Führer Principle and a strong state versus party wrangling and the shilly-shallying of mass democracy. They propose a strong Eurasian community as an alternative to the transatlantic alliance and a decrepit European Union.

The antagonism towards modern liberalism has a long tradition in European intellectual history. Its roots can be found in the Romantic infatuation with nature and the contempt for the modern mass society, the advocates of a 'conservative revolution' in the Weimar Republic, the masterminds behind Italian fascism and the Nouvelle Droite in France. Its offshoots can be seen all over Europe.

This is not an academic debate. We are confronted by a struggle for ideological hegemony. It has an uncompromising power-political dimension. It is about supremacy in Europe, about 'Eurasia' as an alternative project to 'Atlanticism'. The opponents of the West speculate on the implosion of the EU and the end of the transatlantic alliance. Dmitry Kiselyov, chief propagandist of Russian state television, cheered the Brexit referendum as a sign of the collapse of the European Union. Even before the referendum, Russian media painted a picture of a declining Europe attacked by hordes of Muslim migrants, abandoning Christian values and indulging in sexual decadence. The opponents of the West speculate that Brexit will herald the end of Atlantic Europe and open the way for a 'united Eurasia' led by Russia. This is not merely wishful thinking. It has resonance in Europe, and the Kremlin is investing money and political capital into the development of its European networks. Trump's anti-European policy suits this strategy perfectly. Putin probably can't believe his luck. The Russian interference in the American presidential election is bearing fruit.

Chinese media loyal to the state also sang of the decline of the EU after the British vote. The image of the West is darkened so that the Chinese model can shine in a brighter light. China's authoritarian regime appears as a guarantor of prosperity, stability and determination in an uncertain world. Who would be foolhardy enough to risk these achievements for more democracy and freedom? This message is having the desired effect, not only in China.

Anti-liberal turning point

It is not easy to be optimistic at times like this. The brief decade of democratic change in Europe is over. Anyone who still hopes that tomorrow's world will be a better one is seen as naive or to be simply trivializing the situation. Insecurity is the new underlying pan-European sentiment. This is not merely a quirk of an ageing society. We are living in a time of change, and it is not certain how western democracies will deal with the looming challenges.

The speed and simultaneity of crises and conflicts do not only test our own personal capacity for understanding; they also overwhelm the political system. The potential for chaos is growing, creating a more aggressive mood. Those who feel threatened seek protection by joining the wagon train. The demon of radicalism, which brought our continent to the brink of self-destruction, is returning. Confidence in the competence of the political class and its ability to act is at an all-time low. Populist parties and movements are challenging the basic consensus that the liberal elites in the West had agreed on: an open, tolerant and multicultural society, gender equality, recognition of gays and lesbians, globalization of the economy and the development of transnational institutions.

The constant flow of bad news with which we are bombarded on the internet reinforces the impression that the world is out of joint – at least the world in which we felt more or less secure. Terrorist attacks by

Islamic jihadists introduce a fear of random violence to our everyday lives. The European Union is experiencing the deepest crisis in its history. The supposedly irreversible law of motion towards an 'increasingly close political community' no longer applies; the creeping erosion of the Union is a real possibility. In many European countries, economic stagnation and high youth unemployment are clouding the future prospects of the young generation. The gap between the winners and losers of modernization is growing. The massive flight of people from the war and crisis regions on the other side of the Mediterranean is splitting European societies. Turkey has taken a dramatic turn and is distancing itself rapidly from the West; with Erdoğan, it is on the way to becoming a Russian-style democracy headed by an authoritarian leader. NATO's south-eastern flank is crumbling, while Putin's Russia is once again appearing as the adversary of the West. The European security architecture confirmed in the treaties of Helsinki and Paris – renunciation of violence, territorial integrity and equal political sovereignty of all states – has been effectively invalidated by the annexation of Crimea and the intervention in eastern Ukraine. The Arab world is being shaken by excesses of violence that we are unable to respond to. In the South China Sea, a major geopolitical conflict is developing between a brazen and self-confident China and the United States. A world in which the strategic rivalries of the twentieth century are contested with the weapons of the twenty-first century is a highly dangerous place.

And, as if that were not enough, the transatlantic alliance is creaking at the joints. For Trump, the EU is no longer a strategic partner. He relies on confrontation rather than cooperation. In Europe as well, forces that preach national egoism and have lost all understanding of the West's historical project are on the rise. There is no shortage of voices who are pleased to see the

United States crumbling as a world power. Is it not high time for a new, multipolar world order in which the countries of Asia, Africa and Latin America take their place as equal players on the world stage? No doubt this process is overdue and inevitable. But no one should underestimate the risks and potential for violence associated with the shift in the international balance of power. In the past, the transatlantic alliance provided an anchor of stability in the international system. A reform of this system is overdue, but its collapse would be extremely dangerous. This applies even more to the normative pillars of the international order: the Universal Declaration of Human Rights and international law. To dismiss them as relics of western hegemony would be a dramatic setback in the development of a global civilization. A just world order requires a normative foundation.

2

Modern and Anti-Modern

The history of modernity has been accompanied from the outset by anti-modern movements. 'Modernity' describes an era of rapid social change initiated by a combination of philosophical enlightenment, scientific and technological revolution, and the rise of democracy and human rights. This stimulus is still active today and has even received a further boost through globalization. All the declarations pronouncing the death of modernity were premature. We have not entered a new era of postmodernity but rather a phase of *global modernity*. It is not a mere copy of the classical modernity formed in Europe and America but rather is producing a pluralistic and diverse modernity.

Modernity has never been a 'one-size-fits-all' phenomenon, but the different forms nevertheless share certain features: the transformation from agrarian to industrial societies was accompanied by rapid urbanization, higher levels of education, social and geographical mobility, distinct living styles and the growth of the middle classes. It is unclear whether this will also apply in the long term to the democratic side of modernity, the striving for self-determination and self-government. We can no longer rely on the good old 'stage theory', according to which economic modernization, rising income and education automatically lead to democracy and the rule of law. It seems likely that in the long run a self-assured globally networked middle class will not be satisfied with the material conveniences of modernity but will also demand civil liberties. There are also good reasons for assuming that the transition to an innovation-driven economy will lead to conflicts with centralized power structures and authoritarian hierarchies. Private property demands legal security; complex national economies are reliant on a transparent information flow, market-dependent coordination and a high degree of personal responsibility. But the trend towards democracy is not inevitable. Nationalism and fear of instability are strong counter-forces. The Chinese example will show whether it will be possible in the long run to keep the processes of modernization and democratization separate. It is no exaggeration to say that the future world order will depend on this question.

Escape into the community

Every new stage of social modernization engenders fear and resistance. There is a recurrent dialectic between modern and anti-modern tendencies. Scientific demystification of the world versus romanticism, secularization versus religious fundamentalism, individualism versus the longing for community, globalization versus nationalism, and permanent change versus the desire for security are all opposites occurring again and again in ever-changing forms. Karl Popper described this conflicting constellation as the opposition between 'open' and 'closed' societies. *The Open Society and Its*

Enemies, his monumental study of the history of ideas, was published in London in 1945. Popper wrote at a time when continental Europe was dominated by the two totalitarian powers of the twentieth century. It was his contribution to the struggle for freedom.

Twenty years earlier, Helmuth Plessner had published his far-sighted book *The Limits of Community*, which discussed 'social radicalism' on the left and right.[1] For Plessner, communism and fascism have a common core: combating modern society in the name of the community. This was not a random idea. As a response to the collapse of the empire and the 'dictated peace' of Versailles, a conservative revolutionary movement formed in the Weimar Republic calling for 'national socialism' and a foreign policy based on an anti-western alliance with the Soviet Union. One of its most influential masterminds was Arthur Moeller van den Bruck, whose major work *Das Dritte Reich* (The Third Reich) became a powerful political pamphlet. Moeller's main ideological enemy was liberalism as an individualistic 'community-destroying' theory and way of life: 'Liberalism is the expression of a society that is no longer a community.'[2] For all their political differences – the one *völkisch* nationalist and the other proletarian internationalist – there was a common denominator between the fascist *Volksgemeinschaft* (people's community) and Bolshevism. Both were homogenizing utopias, and it is that which gave them their potential for violence. The homogenization of societies with highly differentiated social, ethnic and cultural components came down to a policy of destruction – of the 'class enemy' on the one hand and of *bluts- und volksfremde Elemente* (non-German and non-ethnic elements) on the other.

Community stands for the immediacy of personal relations, the continuity of traditions and the primacy of the collective over the individual. *Society* is the public sphere, the institutions, the functional relations, the 'art of business' and a game with changing roles. In communist and fascist anti-bourgeois discourse, society is equated with alienation. The primacy of abstraction (law, money, global trade) is contrasted with the utopia of immediacy, nearness and community. In both, it is a case of *eliminating differences* – be it in a classless society or in the fascist *Volksgemeinschaft*. The removal of heterogeneity leads to violence. The elimination of the class enemy – the bourgeoisie, the kulaks, the counter-revolutionary elements – is equivalent to the destruction of the alien race that threatens the national body. It is no coincidence that 'putrefaction' and 'purification' were central battle cries in both Stalinism and Nazism. The road to homogeneity is lined with massacred bodies.

The crucial point is that the opposition to modernity comes from within itself. This conflict penetrates as far as the awareness of individual members of society. Most people have an ambivalent relationship to modernity, albeit to differing degrees. The history of progress has two sides. It is in fact progress towards a richer life and, in the long run, also towards democracy and civility. At the same time, it is a history of losses and horrors of all kinds. Walter Benjamin expressed the negative side of progress in a deeply pessimistic picture: modernity is a whirlwind pulling us backwards into the future, while we look back on a chain of disasters. The ceaseless change makes many people tired. It was Robert Musil who coined the aphorism 'progress would be wonderful, if only it would stop.'

Two world wars, the Holocaust and the atomic bomb have deprived modernity of its confidence in progress. Its destructive potential is darkening the future. It is true that in many respects modernity is a risky historical formation. Setting free the individual also implies a loss of social cohesion. Global markets are volatile. Science and technology are enormous productive forces but they also harbour potential threats. Competition produces winners and losers. Nothing stays as it is. This produces anxiety and the desire for the support of an established community. It should not be forgotten that fascism and communism, the two major totalitarian counter-movements to modern liberalism, were born in Europe. They were combative alternatives to 'bourgeois democracy', to the individualistic and materialistic culture of the West. They filled the metaphysical emptiness of liberal democracy with a this-worldly religion, which made individuals feel as if they were part of a heroic whole. Ideology- and emotion-driven movements fulfil an elementary psychopolitical function: they respond to 'the desire of men to find and to know their definite place in the world, and to belong to a powerful collective body.'[3]

Alexander Dugin's crusade against modernity

Anyone studying today's masterminds spearheading the opposition to modern liberalism will stumble sooner or later on the Russian politician and publicist Alexander Dugin, the intellectual Rasputin of the New Right who operates at the border between Bolshevism and fascism. He sees 'modernity and its ideological basis – individualism, liberal democracy, capitalism, consumerism and so on' to be the cause of the future catastrophe of humanity, and the global domination of the western lifestyle as the reason for the 'final degradation of the Earth.'[4] Dugin's anti-liberalism rant is in no way original. He joins a long line of nationalist revolutionary ideologues. He copies almost word for word what Moeller van den Bruck wrote in 1923: 'Liberalism has undermined cultures. It has annihilated religions. It has destroyed nations. It is the self-dissolution of humanity.'[5]

He agitates incessantly against the liberal universality of the West, which seeks to impose its way of life, its values and institutions on the entire world – be it through subtle mechanisms of fashion, music and cinema or through brute force. For Dugin, liberalism's claim to global validity is merely a hidden form of totalitarianism. Its passion for freedom is in reality the 'most disgusting formula of slavery' because it deludes people into rebelling against 'God, against traditional values, against the moral and spiritual foundations of his people and his culture'.[6] Now that liberalism has defeated its ideological opponents – communism and fascism – it is permeating the whole world. The United States and its vassals are doing their utmost to destroy the traditional way of life of other nations and to replace it by the 'operating system of American liberalism'. Those – such as the Taliban in Afghanistan, Saddam Hussein in Iraq or the Serbian ethno-nationalists under Slobodan Milošević – who defend their traditions against the uniform liberalist religion are subdued with bombs and shells.

Dugin thunders against:

- western individualism ('the individual as the measure of all things');
- the 'sanctification' of private property;
- equality of opportunity as a 'moral law of society';
- the commitment of all political institutions to contractual agreements with bourgeois society;
- the abolition of all state and religious authority claiming to represent 'the common truth';
- the separation of powers and the establishment of control mechanisms over every state institution;
- the primacy of 'market relations' over politics; and
- the presumption that 'the historical path of western nations is a universal model'.[7]

All this is part of the standard repertoire of the critics of modernity. Dugin gives it a specific mystical *völkisch* hue. He evokes tradition and religion, nation and territory as the fateful foundation for all existence. He counters modern rationalism with religiously charged politics and pragmatism with heroic gestures: 'either our political struggle is soteriological and eschatological, or it has no meaning.'[8] The final struggle will be against America as the centre of modern liberalism: 'The American Empire should be destroyed. And at some point, it will be.' Dugin's dream is a 'global crusade against the United States, the West, globalization and their political-ideological expression, liberalism.' This battle should be joined by all forces wishing to defend their traditions and values against liberal imperialism: Russia and its Slav brothers, China and the other Asian peoples, the Islamic world and the anti-liberal opposition in the West. Thus the left-wing anti-imperialism of yore is returning as 'red fascism'.[9]

3

The Long View of Democracy

Human rights are by their nature universal. They are western because of their *origins*, not their *validity*. Through their establishment in the Charter of the United Nations, they have become formal universal values as well. Often enough, anti-colonial movements have proclaimed the ideas of freedom, equality and human dignity against those who originated them. In Europe and America as well, they have become the rallying banner for all those social groups that are deprived of them: slaves, factory workers, women in all countries and sexual minorities.

The history of democracy from its beginnings in the classical polis can be read as a progressive expansion of claims to equality and participation. The abolition of slavery and serfdom, universal suffrage, gender equality and the right to sexual difference were all hard-fought battles. On top of the struggle for political equality comes the struggle for social rights, such as the right to education or collective insurance against unemployment and old-age poverty. They form the material basis of freedom. The process of democratic emancipation is not over and probably never will be. It takes a new form in every generation. With the digital revolution comes the issue of digital civil rights; globalization demands new forms of democratic regulation beyond the nation-state; worldwide migration places the question of multi-ethnic democracy on the agenda.

The idea of universal human rights and the self-governance of free citizens first became a constituent of politics in the American Declaration of Independence in 1776: 'We hold these truths to be self-evident, that all men are created equal, that they are endowed by their Creator with certain unalienable rights, that among these are life, liberty and the pursuit of happiness.' It is the task of the government to protect these rights. If it fails to do so, it forfeits its legitimacy: 'Whenever any form of government becomes destructive to these ends, it is the right of the people to alter or to abolish it, and to institute new government, laying its foundation on such principles.' It is not the state that grants its subjects the various rights; on the contrary, state authority comes from the people. 'People' here is not some mysterious entity but a political category: it is the community of free citizens who claim full political sovereignty.

The French Revolution was the reflection of this unprecedented call for the overthrow of all conditions in which civil rights and liberties were not respected. On 26 August 1789, the National Assembly in Paris announced the Déclaration des Droits de l'Homme et du Citoyen (Declaration of the Rights of Man and of the Citizen), which remains the Magna Carta of modernity. Unlike the American model, it does not appeal to the Creator. Its overriding principle is the freedom of the individual: all men are born free and with equal rights. The goal of any political association – and the democratic state is nothing other than a political association of free citizens – is the conservation of the natural and imprescriptible rights of man.

It mentions explicitly the 'right to liberty, the right to property, the right to security and the right to resistance to oppression'. Individual liberty is limited only by the right to liberty of other members of society. All citizens are equal before the law and have the right to participate themselves or through their representatives in public affairs.

This is still the core of democracy as a political order for freedom. Even today these texts have lost nothing of their critical power. They are an encouragement to everyone seeking freedom and a declaration of war against the authoritarian regimes of our time. And they are an incitement to continue the historical modernity project, the self-determination of the individual and the self-government of society.

The fall of the Berlin Wall and the collapse of the Soviet Union were a triumph for freedom. Nothing remains today of the euphoric mood of those times. When did we give up hope of a new golden age of peace and democracy? It was probably not a single event but a whole series of fatal occurrences that caused things to go awry. In 1991, the Yugoslavian wars of secession began before our eyes with massacres and displacements. Almost at the same time as the massacre of Srebrenica, the genocide in Rwanda resulted in the slaughter of almost one million people. The next major caesura came on 11 September 2001, when radical Islamists carried out their horrific attack on the centre of liberal capitalism. Among the knock-on effects of this attack was America's intervention in Iraq, triggering a continuing escalation of violence throughout the whole region. The financial crisis in 2007 shook to the core the faith in the blessings of globalization. It gave rise to a serious recession, from which Europe has still not recovered completely. And if that were not enough, climate change appeared on the horizon, threatening the livelihood of billions of people. The pictures of overflowing refugee boats and angry young men rattling the European border fences offer a foretaste of future crises and conflicts. It is this accumulation of threatening developments that has turned optimism into insecurity and anger. Confidence in democratic institutions is on the wane. All this lays the foundations for parties playing on fear and anger as they set up a front against liberal democracy. What do we have to counter them except for a defence of a status quo that is being eroded from all sides?

Splendour and misery of liberalism

The situation is paradoxical. For a long time, it looked as if liberalism had won definitively. After the collapse of the Soviet empire, the flag of democracy was hoisted throughout Europe. The European Convention on Human Rights was signed by forty-seven states. Written constitutions guaranteed freedom of opinion and the media, separation of powers and free elections. There did not appear to be any serious alternatives to capitalism. It sounded like the final triumph of liberal ideas. And yet political liberalism in Europe suffered a nasty setback at the same time. With few exceptions, liberal parties had become marginal phenomena that almost no one wanted to have anything to do with. How is this to be interpreted?

One key to explaining this phenomenon is the 2007 financial crisis and its devastating consequences. It broke the back of an enucleated liberalism. Apart from calling for state cutbacks, deregulation and privatization, it had little to offer, and no one wanted to hear this tune any more. 'Neo-liberalism' became the political killer word. It has turned into a synonym for everything that public opinion believes to have gone wrong: untamed finance markets, the primacy of business over politics, turbo-capitalism, growing inequality and social insecurity. Liberalism has been eaten up by neo-liberalism. Today, it is seen by large parts of the population as an ideology used by global elites to camouflage their self-serving economic interests, a code word for uncontrolled enrichment and disdainful disregard for the classes left behind who do not wish to espouse the beautiful new world of open borders, global competition and multicultural diversity.

Barely anyone remembers that neo-liberalism originally stood for the revival of a liberal philosophy and politics. In the years after the Second World War, neo-liberalism was an umbrella term for the attempt to find liberal answers to the aberrations of totalitarianism. At its core it was about questions of an economic constitution and policy. Typically for liberalism, it never took a uniform line. The neo-liberal spectrum ranged from advocates of radical market liberalism to the Freiburg School, which gave rise in Germany to the concept of a social market economy.

The most prominent of the market liberals was Friedrich August von Hayek. Against the historical background of communism and fascism, he called for strict limitation of state intervention in the economic process. Self-regulating markets bundling the knowledge and preferences of a large number of actors were not just a superior instrument for decision making but were also a guarantee of freedom – no political freedom without economic freedom. In retrospect, Hayek's glorification of the market is merely the reverse of the absolutization of the state. He overestimated the inherent rationality of markets and underestimated their destructive tendencies. Political interference in market freedom was seen as a disruption of a self-regulating system.

Together with the American economist Milton Friedman he formed the radical wing of neo-liberalism, which in the 1970s and 1980s became the dominant economic policy ideology. It attacked the Keynesian school of economic policy, rejected debt-financed economic programmes and called for the minimization of state interference in the economic process. To restore economic dynamism, welfare spending should be cut, state-owned businesses privatized and markets deregulated. The growth unleashed in this way would lead to increasing employment and higher earnings, from which the poorer classes would also ultimately benefit.

This was the blueprint for revolution from above imposed by Margaret Thatcher in the 1980s in the United Kingdom against all resistance. She capped subsidies for loss-making industries, privatized state-owned businesses and abolished posts in the civil service. While industry shrank, the City of London rose to become the global financial centre. Private enterprise gave new impetus to the British economy. Financial services, culture, media and tourism became growth industries. The social costs of this radical cure were enormous. Entire regions lost their economic base. Regular employment was replaced by precarious jobs. The social gap became wider as winners and losers of the neo-liberal revolution lived increasingly in different worlds.

Unlike Hayek and Friedman's Chicago boys, the German *ordoliberals*, led by Walter Eucken, put limits on both the state and the market. They sought a third way between laissez-faire capitalism and a planned economy. Hayek was criticized for having done a great disservice to the renewal of liberalism. The angry word 'palaeoliberalism' was coined. The ordoliberal ideology called for an economic order in which free competition was combined with civil liberties. The market economy is dependent on conditions that cannot be inherently guaranteed. They include not only the right to property and freedom of contract but also public goods such as an efficient infrastructure, education and research, and environmental protection. It is the task of the state to provide the political and legal framework in which forces can develop freely. Institutions such as the *Bundeskartellamt* (Federal Cartel Office) and the state Monopolies Commission are designed to prevent the concentration of market power. Whereas Hayek rejected the flexible concept of social justice as the gateway to boundless state interventions, the ordoliberals assigned the state an active sociopolitical role. Where the market produced social aberrations, the government should intervene to correct the distribution of income – through progressive taxation of income and property, for example.

Eucken and the ordoliberals were nevertheless also far from issuing the state general powers to intervene in the economy at will. They believed that state intervention was justified only if it safeguarded the functioning of the market economy and prevented the misuse of power. The government should concentrate in general on issues connected with the *economic constitution* and resist the temptation to intervene selectively in the economic process.

The vacant liberal position

When did a liberal party last propose economic and social policies that could compete with the quality of the reflection of the classic neo-liberal school? A long time ago. The guiding principle of liberalism since the 1980s has been reduced to the basic formula 'the less state, the better'. Beyond the holy trinity of deregulation, privatization and tax cuts, most liberal parties have not had much to offer.

In Germany, the liberal FDP handed over the leadership of civil rights liberalism extensively to the Greens, who fought more vigorously and systematically for minority rights, spoke up for gender equality and against discrimination of homosexuals, and defended the private sphere from additional state surveillance. The Greens also took over from the FDP in terms of a liberal immigration policy, not to mention environmental protection. For a long time the FDP did not show much enthusiasm for education policy – one of the domains of liberal policy based on equal opportunity, social advancement and responsibility. Its supporters dwindled as a result to a hard core of 'higher earners' who applauded the liberal market agenda. Partly through their own fault, the liberals acquired the reputation of a party of social indifference without eyes or ears for the concerns and worries of the broad majority.

The neo-liberal programme marginalized liberalism in Germany and most European countries. The FDP was overtaken by the Greens as the party of the modern middle classes. They are happy to reclaim the liberal heritage for themselves, but their liberalism is half-hearted, focusing only on the civil rights component. The Greens are in their element when fighting for the rights of women, homosexuals or refugees, and for data protection and freedom of information. By contrast, it is hard to find the liberal element in the Green economic and social policies. Here the dominant tendency is to respond to every problem with a law instead of providing the regulatory framework for a market economy. This applies in particular to the environment, the issue at the heart of the Green movement. There is barely any discussion of what a liberal climate and environmental policy should mean. It is precisely this, however, that will decide whether the Greens really are a modern ecoliberal force or whether they adopt a paternalistic approach to ecology that merely irritates citizens and businesses with a succession of new regulations and prohibitions.

The crisis in the centre left

The decline of political liberalism is an international phenomenon. In the United Kingdom, the attempt to give the Labour Party a new social liberal orientation failed. There was talk for some time of 'New Labour' and the 'Third Way'. Anthony Giddens provided the intellectual framework.[1] In the run-up to the European elections in 1999, there was a joint Schröder–Blair paper claiming to be a modernization concept for European social democracy. It spoke a lot about activating social policy, education, innovation and entrepreneurship. Elements of this Third Way politics were taken up by the Red–Green coalition in their Agenda 2010. The end of the story is well known. Schröder lost the election; the unions and a large number of SPD supporters regarded the social and labour policy reforms as a neo-liberal aberration. In the United Kingdom, things were no better. Blair did not have the stature of a Margaret Thatcher, capable of risking everything for his modernization agenda. The PR message was more important than the actual agenda. His support for the Iraq war at the side of George Bush finally put paid to New Labour. Today the party, headed by Jeremy Corbyn, is seeking a more successful future by returning to the traditional left.

The traditional centre-left parties are in difficulties almost everywhere in Europe. Their old class basis, organized industrial labour, is fragmenting. Their voter base is shrinking. Some of their former supporters have turned to right-wing populists. They have increasing difficulty in attracting divergent social milieus that have practically nothing in common any more: on the one hand, the traditional workers, who see themselves threatened by free trade, open borders and immigration; and on the other hand, the well-educated, cosmopolitan classes cultivating a liberal lifestyle. The first group is turning increasingly to xenophobic, national socialist parties, the other is drifting towards the Greens or to the new alternative left parties, such as Podemos in Spain or Syriza in Greece, which are cutting the ground under the feet of social democracy.

The experiments with New Labour and the New Centre also failed because the shift towards social liberal policies was not supported by the traditional social democratic base. This is illustrative of a general dilemma in liberalism: its narrow social base. There is no liberal people's party to be seen – perhaps with the exception of the political platform launched by Macron in France. This will not change as long as liberal politics cannot be brought into line with the demand for social security and participation. At a time when liberal democracy faces massive challenges, there is an even greater need for a combative liberalism that climbs down from the ivory tower inhabited by the cosmopolitan elites.

Carl Schmitt and the myth of homogeneity

If liberalism is the theoretical form of modernity, democracy is its political form, in which equality *and* difference can both evolve. It transforms people with different origins, social status, cultural influences and political convictions into citizens with equal rights and obligations. Carl Schmitt, one of the intellectual gravediggers of the Weimar Republic, turned everything upside down with his statement:

> Every actual democracy rests on the principle that not only are equals equal but unequals will not be treated equally. Democracy requires, therefore, first homogeneity and second – if the need arises – elimination or eradication of heterogeneity. [...] A democracy demonstrates its political power by knowing how to refuse

or keep at bay something foreign and unequal that threatens its homogeneity.[2]

For Schmitt, equality is an empty abstraction as long as it does not refer to a substantive commonality. This 'substance of equality' consists 'above all in membership in a particular nation, in national homogeneity'. To illustrate his thesis, he mentions '[Turkey's] radical expulsion of the Greeks and its ruthless Turkish nationalization of the country' during the formation of the Turkish nation-state. We can assume that he knew of the Armenian genocide. It was the first violent extermination of the ethnically heterogeneous in the twentieth century, a foretaste of what was to come.

According to Schmitt, democracy can 'exclude one part of those governed without ceasing to be a democracy'. The exclusion of slaves from the Greek polis and the refusal to grant citizenship to colonized peoples in the British Empire do not conflict with a democracy's claim to equality. Equal rights are the entitlement only of those who are substantively equal because they belong to a common nation, which he understands not as a political concept but as a mythical community of destiny. It is only through homogeneity of the members of a state that equality among equals can be achieved: 'In democracy there is only the equality of equals and the will of those who belong to the equals.' 'To govern a heterogeneous population without making them citizens' is not therefore in conflict with democracy.

The postulate of universal and equal suffrage, says Schmitt, is no more than a liberal humanitarian stupidity. The idea that 'every adult person, simply as a person, should *eo ipso* be politically equal to every other person . . . is a liberal, not a democratic, idea; it replaces formerly existing democracies, based on a substantial equality and homogeneity, with a democracy of mankind.' Only the natural homogeneity of subjects allows the 'identity of governed and governing', which for Schmitt is the essence of democracy.[3] From here, it is just a small step to the unity of nation, state and Führer.

All of Schmitt's thoughts and energy are aimed at combating liberal democracy. He attacks liberalism on two fronts. First, he rejects the idea of universal human rights that every individual has a claim to as a human being. At the same time, he argues against the procedural understanding of democracy as a system of rights, rules and processes for reconciling difference and sameness. As he doesn't want to be seen as an out-and-out antidemocrat, he appropriates the term democracy to use it against liberalism. His enemy is 'liberal individualism', which he contrasts with the ideal of 'democratic homogeneity'. He rejects liberalism and transforms democracy into the opposite. Elections are not the only form of democratic decision making, he claims. On the contrary, voting in a voting booth is a private act, but the nation exists only in the public sphere. 'The will of the people can be expressed just as well and perhaps better through acclamation, through something taken for granted, an obvious and unchallenged presence, than through the statistical apparatus that has been constructed with such meticulousness in the past fifty years.'

For the constitutional law expert Schmitt, democracy is ultimately not a *form of government* but a mystical unity between government and people. With this sleight of hand the difference between democracy and dictatorship also disappears: 'Bolshevism and fascism by contrast are, like all dictatorships, certainly anti-liberal but not necessarily anti-democratic.' *Dictatorship is democracy*: as an argument, this really is a *salto mortale*. This idea is echoed today in Viktor Orbán's model of an 'anti-liberal democracy', Russia's 'controlled democracy', and Erdoğan's Führer democracy in Turkey. Democracy has become a zombie term: a mere facade concealing authoritarian rule.

Majority rule is not democracy

The modern nation is a political construct. It is defined neither by origins nor by cultural heritage. It is only through cooperation and joint action that a democratic 'we' is created, a *political community* of people of different origins, convictions and social interests. Privately, they differ in terms of social status, income, lifestyle, religious orientation and cultural preferences. What links them as citizens with equal rights and obligations is their membership of a shared democratic space. In the democratic public sphere, they encounter each other as unequal equals.

Democracies are based on a *risky premise*: trust in the power of argument and the judgement capacity of citizens. The arguments presented in public debate should lead to the formation of well-founded decisions. To that extent, democracy is 'belief in the common sense of others'.[4] This sounds a bit reckless and not just since the Brexit vote in the United Kingdom and the election of a president of the United States whom even fellow party members see as a political risk. It was Friedrich Schiller who had the Polish prince Sapieha say: 'Majority? What is it? The majority is madness; Reason has still ranked only with the few. [. . .] 'Twere meet that voices should be weighed, not counted. Sooner or later must the state be wrecked, Where numbers sway and ignorance decides.'[5]

It is clear that even experienced democracies are not immune to fatal decisions. Common sense is not always with the majority, and the contest to win over this majority often results in appeals to emotions rather than political reason. Unlike authoritarian regimes, however, democracy has institutional checks to correct errors as quickly as possible: separation of power, political pluralism, free media and the interplay between the government and the opposition. If they are disabled in favour of a strong leadership, there is no longer any antidote to political despotism and the primacy of particular interests. As we know, in an extreme case, the delusional ideas of a supreme leader can cause a tragedy affecting half of the world.

To translate democracy as 'rule of the people' is correct but also undifferentiated. Without the separation of power, an independent judiciary, political pluralism, a free press and guaranteed civil rights available to the majority, the rule of the people can all too easily be transformed into tyranny. The majority principle (decision making by the majority) is relativized by the rights of minorities guaranteed by the constitution. In fact, the people do not rule directly but delegate their decision making authority to elected representatives. In most democracies, referendums are the exception to the rule. Legislative authority is in the hands of parliament. This form of indirect rule of the people turns politics into a profession and gives rise to a class of career politicians with their own interests. Their dependence on the sovereign people is created through regular elections, where they are subject to the citizens' vote. Added to this is the ongoing communication with associations, initiatives and citizens during the legislative period. The image of a 'remote' parliament has little to do with the everyday reality of most members of parliament. And yet the title of representative of the people is not highly respected. In Germany, the reduction of parliament to an out-of-touch talking shop has an unfortunate tradition. By contrast, plebiscites are regarded as a higher form of democracy because they provide an undistorted expression of the will of the people.

Reforming parliamentarianism

The left has an ambivalent relationship to parliamentarianism. Historically, the labour movement fought for universal suffrage and parliamentary control of the government. For the radical left, however, parliament was just a disguise for the primacy of capital. It was useful at best as a propaganda platform for revolutionary subversion. In their early days, the Greens were also sceptical of parliamentarianism. Parliamentary work was seen as the 'free leg' of green politics, and extra-parliamentary action as its 'engaged leg'. Parliamentary democracy needed to be refined through plebiscitary decision making. Referendums were regarded as a higher, more authentic form of democracy. 'Basis democracy' was the great magical antidote to the supposed remoteness of parliamentary committees. The term is derived from the idea of direct rule of the people: the 'basis' should decide – members of parliament are just the bodies implementing the direct will of the people. There was thus a good deal of sympathy for the idea of an 'imperative mandate' and little understanding for the independence of members of parliament guaranteed by the Basic Law, the *Grundgesetz*, Germany's constitution.

The contrast between liberal and mass democratic ideas can be found on both the right and left sides of the political spectrum. Once again, Carl Schmitt has something to say on the subject. His essay *Die geistesgeschichtliche Lage des heutigen Parlamentarismus* (translated into English as *Crisis of Parliamentary Democracy*), published originally in 1923, is a classic anti-parliamentarian work. Schmitt sought to divorce parliamentarianism from democracy. He saw free and universal suffrage, party pluralism and parliamentary legislation as aspects of the hated liberalism, while he understood democracy as an organic unity of a nation-state and political leadership that did not require parliamentary legitimation. He is nevertheless worth reading. With greater precision than most defenders of parliamentarianism, Schmitt presents in detail the central principles on which it is based. The notion and practice of parliamentarianism is inseparably linked with the republican principle of public debate. It is a medium for democratic opinion forming. Parliament is the centre for this public debate. At best, discussion is not mere shadow-boxing between firmly established party blocs. It is rather 'an exchange of opinion that is governed by the purpose of persuading one's opponent through argument of the truth or justice of something, or allowing oneself to be persuaded of something as true and just.'[6]

Unlike business transactions or wage disputes between unions and employers, it is not a contest between *interests* but between *opinions*. For that reason, parliamentarians should see themselves as representatives not of special interest groups but of the people as a whole. The objects of parliamentary debate are different assessments of the situation, different aims and priorities, and ultimately different ideas of the common good.

Schmitt spells out this idea so as to highlight even more forcefully the gap between it and parliamentary reality. The way he describes normal parliamentary business is all too familiar to us. According to the constitution, members of parliament are free to decide as they wish and are not bound by any instructions. In practice, they are subject to a party whip, who decides who should speak and how its members should vote. Those who want to be re-elected would do well to toe the party line. The result of a vote can usually be predicted before it is announced. The members of the majority party are duty-bound to support the government, while the opposition does its utmost to find fault with it. Debates that are really an attempt to persuade and that make parliament a focus of public attention are rare and notable exceptions.

Schmitt offers a further classical justification for parliamentarianism. Its supporters saw it as 'a means for selecting political leaders, a certain way to overcome political dilettantism and to admit the best and most able to political leadership.' This is followed by a disdainful comment that this ideal has not worked out in practice either:

> What numerous states in various European and non-European states have produced in the way of a political elite of hundreds of successive ministers justifies no optimism. But worse and destroying almost every hope, in a few states, parliamentarianism has already produced a situation in which all public business has become an object of spoils and compromise for the parties and their followers, and politics, far from being the concern of an elite, has become the despised business of a rather dubious class of persons.[7]

In a nutshell, he offers a critique of the parliamentary party system. He attacks parliamentarianism from an authoritarian and *völkisch* perspective. But that does not prevent his critical discussion of the dark side of party democracy, which he uses as ammunition.

Where parties extend their influence to the occupation of public offices, cultural institutions and state-owned business, they need to be reined in. They should concentrate on the task that our constitution has allocated to them: participation in political decision making and giving a direction to public debate. For this to happen, they must become forums for lively debate again. Parties that streamline their policies in accordance with opinion polls, whose conferences are mere PR events and who discourage criticism from within are failing in their task. They should be more open to outsiders from the academic, business and cultural world, instead of encouraging party political incest. It is wrong that most members of the Federal Republic's political class have barely any experience outside the world of party politics.

If parliamentarianism is to regain respect, parliaments must become more interesting again. This is first and foremost an appeal to our elected representatives to take parliament seriously as a platform for democratic debate – no more prepared speeches, more live debate, less party political ritual, more cross-party motions and non-aligned voting. The government should be monitored by parliament as a whole and not just by the opposition. Ministers should become used to hearing critical questions from the ranks of the majority parties as well. Similarly, it does nothing for the opposition's reputation when it automatically condemns every government decision.

Parliamentarians must have the courage to follow their personal convictions and, if necessary, risk conflict with their own party. Members of parliament who always follow the directives of the party leadership are not free but mere foot soldiers.

Democracy is the home of freedom

The greatest danger to democracy comes from within. It grows weaker as soon as we stop appreciating it as the guarantor of freedom. If freedom is the *raison d'être* of politics,[8] democracy is the home of freedom. It is true that security is important: security from despotism and violence, and security from want are conditions for freedom in action. But those who make security the supreme purpose of the state sweep aside democracy. 'Security above all' is the false promise of all autocratic rulers, the siren song with which enemies of the open society seek to capture votes. They call for reducing freedom in the name of combating terrorism, stemming the massive flow of immigrants and protecting against the constant changes taking place in the modern world. The defenders of liberal democracy should not ignore the need for security, but they must be clear in their purpose when it comes to defending freedom from the obsession with security.

Those who play off security against freedom will ultimately lose both. This message is aimed not only at those who support a 'security state'. It also applies to lovers of freedom. Whenever social security is the subject, the political left feels addressed. Social democracy even has the link between participation in society and democracy in its name. It is a different matter when it comes to the classical issues of internal and external security. The liberal left is right to worry about the degeneration of democracy into a police state and militarism. They are making a great error, however, if they leave the area of security policy to the conservatives. We no longer live in the era of the imperial authoritarian state but in a democratic republic.

The democratic state is the state of its citizens; it is authorized by us and protects our freedom. *Defending freedom means defending the democratic constitutional state.* Classic liberalism, which arose in opposition to the absolutist state, was concerned above all with the threat to civil liberties by state authority. This issue has still not been solved today. We should not give free rein to the security apparatus. It would be extremely naive to ignore the fact that liberal democracy is threatened by forces from within and without that have a completely different notion of state and society.

Those wishing to defend democracy should be clear about who their enemies are. When Karl Popper published his polemic in defence of freedom at the end of the Second World War, he entitled it *The Open*

Society and Its Enemies. It was evident who these enemies were at the time: the hostile brothers fascism and communism. Today, we recoil at the word. To speak of enemies is to return to pre-democratic thinking. For liberal ears, it sounds too much like Carl Schmitt, who saw the essence of politics in the distinction between friend and foe. The scepticism is justified. If the dualism of friend and foe is raised to a political level, politics becomes merely a continuation of war by other means. With this logic, all internal policy becomes latent civil war and all foreign policy a latent war between states.

The transformation of opponents into enemies is rooted in the struggle for power. Democracy is not immune to this temptation either. It flares up above all in election campaigns. It includes the demonization of the opponent and the elevation of political differences within the democratic spectrum to a struggle between good and evil. Parties based on the constitution should not treat each other as enemies. Between enemies, the question is always who will be victorious in the end. Between political opponents, compromises are possible. In a parliamentary republic, the contest for a majority must be tempered by an awareness that the foundations for constructive cooperation should not be destroyed.

The fact of being cautious in categorizing enemies does not mean that democracy has no enemies. Right-wing extremists, nationalist movements, radical Islamists and left-wing extremists are out to overthrow the democratic constitution. Should we apply to them the maxim 'no freedom for enemies of freedom', or the principle that freedom is always the freedom to think differently? Germany's constitution makes it possible to ban political associations seeking to overthrow the democratic order. The conditions it sets are rightly strict. But liberal democracy is also inevitably confronted by the distinction between friend and foe. The memory of the overthrow of the Weimar Republic by the

Nazis remained alive in Germany until the 1970s. The emergency laws in 1968 (under a coalition of the CDU/CSU and SPD), and the employment ban for communist cadres in the 1970s under a social liberal coalition with Willy Brandt as federal chancellor may be criticized as panic overreactions. They can be understood only against the background of the experience of a generation who concluded in the wake of the Weimar history that democracy must defend itself against its enemies.

Militant democracy

This lesson has become relevant again today. My generation grew up with the paradoxical belief that the greatest danger to democracy comes from the democratic state. The defence of freedom was directed in the first instance against state control and surveillance. The mass hysteria that broke out in 1987 in Germany against the planned census revealed the deep-seated mistrust of the integrity of democratic institutions. Some members of the public invoked the idea of an out-and-out police state. As so often in Germany, it was a case of subservient obedience or hyperventilating total refusal. This appears still to be the basic pattern of public arousal, from the debate on data retention to the mass protest against the free trade agreement with the United States and Canada.

A good proportion of civil liberties are rights protecting against state absolutism. They are designed to rein in state authority, to protect the private sphere and to defend democratic freedoms from interference by state bodies. This is as it should be. Protection from government despotism is a precious achievement. It was wrested from absolute rulers step by step. Even in a democratic state governed by the rule of law, civil liberties are never secured for all time against intervention by the state. As we have seen recently, there is always a possibility that democracies will fall into the hands of anti-democratic governments. By allowing the state the monopoly on legitimate violence, we give it wide-ranging instruments of power. Those who have power can abuse it. In many countries, the state authorities are indeed the greatest threat to freedom. This applies where there is no functioning rule of law, and criticism from the public as a controlling instance is suppressed. Even within the European Union, elected governments are imposing their authoritarian will on society, as in Hungary and Poland. There is a need to remain watchful of the threat to freedom from those who are meant to protect it.

It is nevertheless a distortion to see the threat to freedom primarily in the institutions of the democratic state, while at the same time we are confronted by actors who make no bones about their disdain for liberal democracy. They use the political freedoms that the liberal state governed by the rule of law also offers its opponents to attack it. In Russia, critical citizen initiatives are stigmatized as 'foreign agents', opposition media are pressed up against the wall, and foreign foundations are declared to be 'undesirable organizations'. This doesn't prevent the Kremlin from investing a lot of money into the activities of Russian media and foundations in Europe and from developing its political networks in the West. Militant Islamists invoke freedom of opinion and religion while at the same time cultivating links with ISIS, al-Qaida, Hezbollah and Hamas. Extreme right-wing parties and movements are networked throughout Europe and use the EU parliament as a political platform.

Defending freedom calls for a watchful civil society, a critical public and the systematic application of law and justice. A strong democracy can exist only when civil society and the rule of law work together.

We can defend the rule of law only with its own instruments. Opponents of democracy also have rights that must be respected. We can argue whether political hate tirades can also claim the right to freedom of expression, at what point anti-constitutional organizations should be banned, and where the limits of academic freedom lie. In general, the room for public discussion of opinions should be as large as possible. As long as they do not overstep the boundaries of incitement of the masses, expressions of opinion, even the most audacious, should not be treated as criminal offences.

The central line of defence of democracy is the *non-violence* of political discussion. There can be no concessions here. Those who use violence to prevent others from exercising their rights are killing democracy. The radical offshoots of the student movement in 1968 rejected the state monopoly on violence and proclaimed the right to violent resistance. They simply ignored the difference between the democratic state, whose monopoly on violence is bound by the law, and a dictatorship, against which violent resistance might be legitimate when the possibilities for non-violent change are cut off. The right to *civil resistance* to democratically legitimized decisions can also be claimed. But the right to counter-violence exists only against a *coup d'état* aimed at abolishing democracy. Playing with violence is playing with fire. Even in small doses, the use of political violence destroys the democratic society. It is not possible any more to go to demonstrations and rallies without fear, if there is a danger that things will get out of hand. In the end, the public space is taken over by militants from the right and left, living out their civil war fantasies. On closer consideration, the state monopoly on violence is the condition for the political freedom of all. Those who wish to defend democracy should not undermine it.

The concept of *wehrhafte Demokratie* (militant democracy) forces us to rethink our own attitude to state security agencies. It is an old mistake by the left to leave internal and external security to the conservatives. The military, police and security services are regarded in left-wing milieu not as instruments of democracy but as latent threats to it, whose authority and budget need to be reduced as much as possible. If the democratic constitutional state is the political form society adopts to protect its freedom, then it is the state we want. Its security agencies should serve to defend rights and freedom. If they fail to follow this guiding principle, they deserve to be criticized and changed.

The challenge is to commit the security apparatus in democracy to the letter and spirit of the constitution and to subject it to parliamentary control worthy of the name. The compromises between risk prevention and data protection have to be renegotiated as required. In any event, we still have to decide at which point the defence of democracy may be considered to have overreached itself and become a danger to the freedom it is meant to protect.

Democracy is based on the belief that the future is not preordained. It is open to possibilities for shaping it. We have the freedom to start over and change the way things are running. This applies to individual self-determination and to political communities. The properties that the autocrats of this world regard as weaknesses of western democracy – the notorious penchant for self-criticism, the urge to compromise – are in fact their strengths. And it is on this that we can build.

4

The Left and Democracy

In its unshakeable self-esteem, the intellectual and political left stands for everything beautiful, good and true. And of course it regards itself as the custodian of democracy. And yet the association of the left and democracy is by no means self-evident, as a more detailed look at the history of revolutionary theory and politics will show. The effects of the left's disdain for 'bourgeois democracy' can still be felt today. This is particularly true of the Marxist theory of the primacy of economics over politics. In its popular version, democratically elected governments are mere agents of capital, or optionally finance markets or multinational companies. For the traditional left, everything is a matter of the economic system. The question of democracy is secondary. Meanwhile, the experience of the last century has taught us one thing: the overarching importance of democracy and human rights as a safeguard against barbarity. The systemic question today is not capitalism or socialism, but democracy or authoritarian rule.

From anti-authoritarian revolt to disdain for democracy

In the European left, there are two main streams: the liberal stream aiming at free self-determination of all and the expansion of democratic participation; and an authoritarian collectivist tradition that places opposition to capitalism if necessary higher than democracy. For Lenin, 'bourgeois democracy' was merely a device for capitalist class domination. The seizure of power by the Russian Bolsheviks in the name of the 'dictatorship of the proletariat' suspended all democratic fundamental rights. All opposition was eliminated. The socialist end sanctified the dictatorial means. This attitude characterized all communist parties. Participation in parliamentary elections was at best a tactical means to soften up the democratic republic from within. Even in the end phase of the Weimar Republic, the German Communist Party (KPD) was unable to bestir itself to defend the republic. The struggle against Nazism was subordinated to the struggle for the socialist revolution. Its 'principal enemy' was not the Nazis but the Social Democrats, denounced as 'social fascists'. In reality the KPD and Nazi Party were competing for power in the wreckage of the Weimar Republic.

By contrast, the 1968 student movement began explicitly as a revolt against authority. Why did anti-capitalism become more important than democracy for a wide section of the protest movement in the following years? Part of the answer lies in the identification of capitalism with fascism proclaimed by large parts of the movement. 'Capitalism leads to fascism, capitalism must be abolished!' was a popular slogan. As a young revolutionary, I shouted this slogan on more than one occasion. The 'repressive democracy' of West Germany was interpreted as a crypto-fascist intermezzo, the emergency laws of May 1968 as preparation for open dictatorship. Added to the suspicion of fascism came the 'struggle against the imperialist war'. Did the Vietnam War not once again confirm the naked barbarity lurking behind the democratic facade of capitalism? The tradition was evoked of the Paris Commune, the Communist International and the Third World freedom movements; support was shown for Che Guevara, the Vietcong, and the Chinese Cultural Revolution. All these movements had nothing to do with 'bourgeois democracy'. To be a liberal was a term of abuse, often in the even more negative form of 'Scheissliberal'. Liberals were said to have no political backbone, to defend the prevailing conditions and merely to give a democratic veneer to capitalism.

In contrast to the Anglo-Saxon tradition, the idea of liberal democracy was barely established in continental Europe. Its brief spring during the bourgeois revolutions in the mid-nineteenth century ended in the restoration of the authoritarian state. The democratic constitutional states that followed the collapse of the old order in 1918 did not have a stable foundation. The democratic centre was weak; the party landscape was dominated by ultra-conservative, fascist and left-wing radical movements. The political passions and heroic ideas were to be found in the anti-bourgeois forces on the left and right. This continued in 1968. The intellectual beacons that the opposition students were guided by – Herbert Marcuse, Jean-Paul Sartre, Ernst Bloch, initially also Theodor W. Adorno and the Frankfurt School – were masters at criticizing capitalism. This was even truer of the popular political icons, from Rosa Luxemburg to Che Guevara. By contrast, liberal theoreticians like Karl Popper were outsiders at best, if they were not simply discounted as apologists for the capitalist system.

The year 1968 had many faces. The by-products of the protest movement scattered to all points of the compass: hippies and life reformers, Maoists and the orthodox left, feminist groups, civic initiatives, Third World projects, pacifists and militants. Some of the movement drifted into the deluded world of armed struggle and a trail of blood. All means appeared justified in the struggle against the 'capitalist beast'. It is no coincidence that the most toxic flowers of violence bloomed in two post-fascist countries. Germany had a long anti-liberal and anti-parliamentarian tradition. In Italy, politics, even in the post-war period, was a continuation of the civil war by other means. In both countries, the militant left had a general suspicion of democratic institutions and styled itself in the tradition of the 'anti-fascist struggle'. All means were justified to prevent the return of fascism.

The 'red terror' of the 1970s seems today like a bizarre aberration. In fact, the boundaries between it and other factions of the radical left were blurred, also in my personal surroundings. It would nevertheless be absurd in retrospect to condemn the entire '1968 movement'. For millions of people, '1968' remained a great awakening. While some barricaded themselves in revolutionary cells and communist cadre parties, others founded anti-authoritarian children's shops, reform schools, alternative publishing companies and newspapers, free theatres, self-governed businesses, women's shelters and civil initiatives, campaigned for humane psychiatry or set off on the long march through party politics and parliaments.

The discovery of everyday politics, the practical improvement of society through concrete projects, the passion for political debate, the stubborn advocacy of self-determination, equality and democratic participation are perhaps what remains from the trials and tribulations of 1968. That's quite an achievement.

The legacy of Marxism

Anyone studying the sources of anti-liberal thinking cannot escape looking at Marx and the school of thought influenced by him. Even if there are no longer any masses on the Old Continent rallying behind the red flags, remnants of Marxist theory and politics have deeply infiltrated the collective conscience. How could it be otherwise? In its various forms, socialism/communism was one of the most influential political ideas in Europe for more than 250 years. The young and rebellious inevitably drifted towards the socialist camp. Sympathy for socialist ideas was widespread among intellectuals and artists. During the years of Nazi rule, the Soviet Union appeared to many to be the only alternative. On the map of Europe during the Second World War, democracies had dwindled to a marginal existence. At the end of the war, Soviet-style 'real socialism' ruled half of the continent. In the late 1960s, Marxism experienced a renaissance in Western Europe. Marx, Lenin and Mao Zedong became the ideological points of reference of an entire generation of intellectuals. Until the collapse of the Soviet empire, 'scientific socialism' was the official ideology of a huge territory from Warsaw to Vladivostok.

After Christianity, no theory has had such a sustained influence on the thinking of millions of people in Europe. The comparison is not coincidental. Even if it wasn't Marx's intention, communism developed into a this-worldly religion. For his disciples, it was nothing less than a matter of redemption in the here and now, the liberation at last from exploitation and suppression, the elimination of an alienating existence, the reconciliation of man and nature. Never was the discrepancy between high-flying ideals and brutal reality so great as in communism. In practice, it became a totalitarian system of government. The utopia ended in unrestrained violence. After the physical elimination of 'class enemies' and all potential opposition, the terror gradually ebbed. What remained was the centralization of power in the hands of a small clique, the enforced conformity of society, a perfect surveillance regime, the systematic perversion of truth and lies and the brutality meted out to all regime critics.

Unrepentant supporters of communist teaching endeavour even today to pass off the failure of this major experiment as a mere accident of history. The pure theory needs to be rescued from the sobering practice. This whitewashing is intellectually disingenuous. Marxism cannot be saved, even if Marx is not responsible for the crimes committed in his name. Of course, there is no linear connection between theory and practice, not least as Marx stubbornly refused to offer a road map for the 'development of socialism'. He contented himself with inferring the need for socialist revolution from an analysis of the 'laws of motion' of capitalism. Moreover, he formulated only a few pointers for a revolutionary action programme on day X. The fact that socialism came to power in Russia of all places, a feudal agrarian country with rudimentary industry and a marginal working class, turns the Marxist theory on its head. Everything that happened afterwards was due to the attempt to retain power by a revolutionary avant-garde that was determined to reshape the country completely according to its ideas, cost what it might.

Revolution versus democracy

For Marx, the economic rules of capitalism were the driving force behind modern history. Its rise and fall were a necessity of nature. Any attempt to fundamentally reform capitalism was doomed to failure. The economic logic of capital penetrated every fibre of the state apparatus. 'The executive of the modern state is but a committee for managing the common affairs of the whole bourgeoisie,' he wrote in the *Communist Manifesto*. It was not politics that determined the economy but the economic system that determined politics. For that reason, any radical improvement would have to start by overcoming the capitalist conditions of production. Elections, parliaments and the judiciary in Marxist theory were merely superstructure phenomena. The decisive element was the economic basis, which placed its imprint on all institutions of 'bourgeois democracy'. It is no coincidence that 'reformism' was a swear word in communist tradition; reform policy was seen as pointless tinkering with the symptoms of incurable conditions of production. Only revolution could break through the 'dull compulsion of relations'.

The disdain for 'fake parliamentary democracy', the dismissal of democratically elected governments as marionettes of capital, the contemptuous attitude to reforms as 'tinkering with the symptoms', the exaggerated focus on the economic system, the bourgeois self-loathing and the romanticization of the revolution all continue to haunt us. It would be unreasonable to make Marx responsible for the trials and tribulations of later generations. But the fatal devaluation of 'bourgeois democracy' that permeates the history of revolutionary socialism started with him. If not immediately declared to be a mere fiction, it is at best a means to the socialist end. It is the revolution that opens the door to the realm of freedom.

Before the state as a ruling machine became 'extinct' in the classless society, the concentration of all instruments of power in the hands of the revolutionaries was necessary to defeat the resistance of the bourgeoisie and ensure the transition to socialism. In practice, the temporary power monopoly became permanent and the 'dictatorship of the proletariat' became a dictatorship of the functionary class. Marx was blind to the fact that it was not capitalism but totalitarian state power that was the greatest threat to freedom. For orthodox Marxists, democracy was subordinate to the question of the economic system. This proved to be a momentous mistake.

The civilization, democratization and humanization of capitalism appeared impossible to Marx and his community. His student Rosa Luxemburg summed up this view in the phrase 'socialism or barbarity' and saw no other historical options. In view of the horrors of the First World War, which appeared to mark the decline of the bourgeois world, there was some validity in this view. But this simplification was also at the root of the tragic error in 1918 when the leaders of the Spartakusbund called for a rebellion in the name of the socialist revolution against the recently established Weimar Republic. In the eyes of the revolutionaries, the democratic republic was a model with no value, a mere pretext for capitalist class rule. The only thing that counted was the questioning of the system: if the entire system was wrong, then everything was wrong. All that remained was complete rejection or complete revolt.

Adorno summed up this notion in the sentence 'Wrong life cannot be lived rightly.'[1] In the original version, he wrote: 'it is not possible to live properly in private life.' After the experience of Nazism it was impossible for Adorno to lead a successful purely private life. One could interpret this as a call for participation in public affairs. Private happiness cannot be decoupled in the long run from the social situation. In the meantime, Adorno's aphorism has become a popular phrase signifying rejection of the 'system'.

In the founding period of the Greens, the topos 'system opposition' (alias 'fundamental opposition') resurfaced along with talk of *Systemparteien* (regime parties), which members of the Greens did not under any circumstances want to associate themselves with. What exactly was meant by the 'prevailing system' was never clarified. The issue was still ultimately capitalism, which was equated with destruction of nature, with militarism and with exploitation of the Third World. Those who pursued realpolitik, in other words the improvement in practice of society and the economy, were regarded by supporters of system opposition as whitewashers and opportunists. Making compromises bordered on treason. The German Greens coined an apt word for it: *eingeknickt* (caving in). This was tantamount to capitulation. Assuming government responsibility was condemned as abandoning noble principles for the sake of collaboration, which was seen as despicable. Disdain for the policy of small steps, opposition as the only truth, rejection rather than creation was the order of the day, 'be sand not oil in the works!' It shouldn't be thought that this type of revolutionary kitsch disappeared with the transformation of the Greens into a reform and government party. It evidently comes back with every generation.

Like the parliamentary republic, in the Marxist tradition the rule of law is also merely a facade for capital rule. Equal rights and freedoms are destroyed by real economically driven inequalities. This is more important than formal equality because it influences the real life of people. The rule of law conceals the real inequality under the veil of formal equality. The writer Anatole France described this paradox eloquently: 'The law, in its majestic equality, forbids the rich as well as the poor to sleep under bridges, beg in the streets and steal bread.' The rich have not only a comfortable life but also more opportunities for asserting their interests politically: through personal networks, access to decision makers and, not least, through the dependence of politics on the economy. Anyone who can threaten to shift his investments abroad or close a factory obviously has more political clout than the man in the street – hence the popular expression, 'Wer das Geld hat, hat die Macht – und wer die Macht hat, hat das Recht' (money gives power and power gives legitimacy).

The talk of 'fake democracy'

The key point in Marxist criticism of democracy is the differentiation between formal and material freedom. It translates into the contrast between formal and real democracy. The parliamentary republic is a fake democracy as long as it rests on the foundation of capitalism. 'Real democracy' can only be achieved beyond capitalism, when economic inequality has been finally eliminated. Although almost no one today calls for the 'dictatorship of the proletariat', remnants of this once powerful school of thought are still very much alive. The belief that parliaments and governments are merely Potemkin-like covers for the 'primacy of business', and that politics is dominated by 'international finance capital', is hard to shake off. It is regularly nourished by both the extreme right and the extreme left. Does reality not continuously provide new material to demonstrate such theses? The discrepancy between legal equality and social inequality and the dependence of democratic politics on a prospering economy are undeniable. Economic crises, rising unemployment and declining tax revenue are dangerous for every government but particularly for those that are subject to free elections. When insolvent banks threaten to ruin entire national economies, governments intervene with taxpayers' money to prevent a chain reaction. The influence of business lobby groups on legislation cannot be overlooked. It is no secret that businesses attempt to influence government actions through all possible channels. The CEOs of large companies can generally obtain a hearing from ministers and heads of government.

The decisive question, however, is whether it follows from the discrepancy between legal and political equality and socio-economic inequality that 'bourgeois democracy' is just Scotch mist. Is it possible to speak seriously of 'fake democracy' just because elected governments do not have the freedom to pursue their policies without the economic repercussions that these involve and must therefore take account of 'the markets'? The opposite is true: it is democracy that offers the working classes the opportunity to assert their interests in the face of the self-driven capitalist economy. The equality of different people, equal rights as citizens and their equality before the law are among the greatest achievements of mankind. At the same time, free and universal elections, the right to union organization, freedom of expression and the media are powerful levers for civilizing capitalism. The gradual expansion of democracy in Europe in the nineteenth and twentieth centuries was accompanied by an unimagined social advancement of the working class, which Marx believed would be possible only in a remote communist future. The labour movement, armed with the right to strike and to vote, made possible a hitherto unknown extension of social rights and achievements. The rapid rise in the level of education and income has led to the creation of a large middle class.

5

The Rise of the Anti-Liberals

When the Berlin Wall fell and the urge for freedom brought down one authoritarian regime after another, it appeared for a historical moment as if a golden age of democracy and peace was beginning. The rivalry between capitalism and socialism was a thing of the past. It seemed only a matter of time before the duo of democracy and market economy would reach the farthest corners of the globe. No alternative systems any more, no ideological battles, just varieties of a single universal paradigm.

That was long ago. Today, there is no question any more of a global victory of liberal democracy. On the contrary, it is under enormous pressure from within and without. In Europe, anti-liberal parties and movements are on the rise. The Front National is the second strongest party in France. In Austria, the Freedom Party (FPÖ) has overtaken the Social Democrats. Among the working class it has an absolute majority, and among young men it is by far the strongest party. The survivors of the old industrial society are protesting against the post-industrial upper class: against the glitterati and expense account society but also against the environmentally and health conscious LOHAS demographic, who look down disdainfully on the pleasures of the lower classes: a car of their own, a schnitzel and a package holiday in Mallorca.

In Finland, Sweden and Denmark as well – among the most prosperous and egalitarian societies in the world – xenophobic parties are on the rise, and also in Belgium and the Netherlands. This list on its own indicates that the ascendency of anti-liberal protest parties cannot be adequately explained by socio-economic reasons alone. Apart from the actual or feared social *déclassement*, there are other factors that spur on the populists: the identification with a nation defined by origins and tradition; the invocation of a threatened national culture; opposition to mass immigration; the call for a strong state as the bulwark of the little man. Then there is the growing gap between the self-referential elite discourse and the everyday reality of those who are lost in the brave new world of globalization, multicultural society, gender mainstreaming and same-sex marriage.

In Slovakia and Hungary, in Bulgaria and Greece, neo-fascist movements have gained seats in parliament. Hungary, Poland and Austria are ruled by national conservative forces, and Italy has a coalition of a populist anti-establishment party and an extreme right-wing formation. They all deliberately oppose post-nationalist and multicultural ideas. Hungary's strongman Viktor Orbán proposes the ideal of 'illiberal democracy', in which the nation is the most important element. In Poland, Jarosław Kaczyński sees the election victory of his party as authorizing him to bring the constitutional court and public radio under his control. He wants nothing to do with a Europe of immigrants, atheists, equal opportunities officers or gay parades. Poland's foreign minister Witold Waszczykowski summed up the discontent with modernity: 'As if the world had to move . . . in only one direction: towards a mixture of cultures and races, a world of cyclists and vegetarians, who only use renewable energy sources and combat all forms of religion. This has nothing in common with traditional Polish values.'

The Polish national conservatives see their election victory as a mandate for defending 'traditional values' – fatherland, family, church – against the aberrations of a misdirected modernization. They regard themselves as the enforcers of the true popular will, which has priority over any constitutional court. In the Polish parliament, Kornel Morawiecki stated: 'Law is important, but law is not sacred. The good of the nation is above the law.'[1] The good of the nation also apparently demands that all exceptions to the already strict ban on abortion should be removed. With the blessing of the Catholic bishops, abortion should be prosecuted as a criminal offence, even in the case of rape or of danger to the life and limb of the mother. The celebrated Polish desire for freedom is reduced to a patriotic right of resistance to the 'diktat of Eurocrats'. In case of doubt, the ruling populists give national self-determination priority over European norms. In the eyes of right-wing patriots, the European Union is at best a necessary evil. Under no circumstances do they accept the primacy of European institutions and governments over the national 'will of the people'.

Light and reft

> *manche meinen/lechts und rinks/kann man nicht velwechsern/werch ein illtum!* [some say light and reft / cannot be foncused / wham a tistake!]
>
> Ernst Jandl

The anti-liberals find an audience not only among the *déclassé* and marginalized. They are successful in both economically struggling and prosperous countries. They also attract petty bourgeois and intellectuals, craftsmen and lawyers. Their mood swings between status angst and rage. They feel ignored by the established politicians and unnoticed by the mainstream media. It is not only or primarily about social issues. A growing number of people are dissatisfied with the way things are going in general. They used to be part of the silent minority, but now they are emerging from the shadows and making a din. They are driven by anxious men who don't really know what is happening to them or where it will all end. The disintegration of the patriarchy has released much destructive energy.

In Germany, the democratic demarcation line keeping out *völkisch* nationalist movements was successfully defended for decades. The eloquent statement by Franz Josef Strauss that no party should be established to the right of the CSU held sway. Election successes by national conservative or radical right-wing formations were nine-day wonders. This barrier now seems to have been removed. If it doesn't self-destruct, the Alternative für Deutschland (AfD) has a good chance of establishing itself in the political landscape. Like all populist parties, its members style themselves as taboo breakers who call a spade a spade and give a voice to the silent minority. They profile themselves as a counterweight to the 'regime parties' and as challenging the establishment. In parliament, they function not as constructive opposition but as aggressors. They fuel the fear of 'foreign infiltration', appeal to social envy with regard to refugees and cultivate the resentment of Germany as the European paymaster. With these arguments, they find support from all parts of the political spectrum. They poach voters not only from the national conservative faction of the CSU but also from traditional SPD and left-wing party voters. The AfD finds greatest support among the working class and the unemployed. Xenophobic rhetoric is particularly effective among those who already feel neglected and disadvantaged. Since its establishment, the AfD has moved ideologically further to the right. It also borrows issues from the left-wing spectrum, however.[2] The attack on the 'arrogant elites', the appeal to the true will of the people and the demand for direct democracy, the trend towards protectionism, the anti-American sentiment and the affection for Putin are features of both the left and the right.

The linking of traditionally left-wing and right-wing positions is particularly marked in some related European parties such as the Front National. Le Pen and the rest present themselves unabashedly as advocates of the 'little man', who demands a strong state to protect him from the assault of globalization. They oppose open markets, demand the protection of families, pensioners and small businesses, fulminate against 'neo-liberal austerity policies' and promise to protect jobs from competition by migrants. An active immigration policy is seen as a neo-liberal plot to reduce wages and social standards – an argument that resonates as well with left-wing populists such as the former SPD chairman Oskar Lafontaine, or Sahra Wagenknecht, the figurehead of the socialist left. Anti-capitalist noises can also be heard from the AfD camp. Attacking international finance capital is not a prerogative of the left. It is easily forgotten that European fascism was also an amalgam of nationalist and socialist ideas. The designation 'National Socialist' in Germany was meant in all seriousness.

The new demarcation line in European politics no longer runs between left and right but between an open society and the withdrawal into a national community, between global integration and national isolation. The old left–right axis is overlaid by the new line of conflict between national authoritarian and liberal democratic politics.

In many respects, today's Europe is a continent that fears for the future. It fears Islam, the migration of the young and hungry from Africa and the Middle East, competition from aspiring industrial countries, the replacement of humans by intelligent machines, gene technology and chlorine-washed chickens, nuclear energy and climate change, pervasive surveillance and the loss of privacy, old-age poverty and the shortage of care personnel. The list goes on.

Some of these concerns have a realistic basis. There is no room any more for a blind belief in progress that ignores the risks and side effects of fundamental changes. The problem begins where technical, economic and social changes are seen primarily as dangers. When we no longer believe that present-day challenges can be met with modern means – science, innovation and democracy – the aggregate political situation of our societies also changes. Then regressive attitudes have the upper hand: isolation, defence of the social status quo against other groups, nationalism, a sharp distinction between 'us' and 'them', mistrust, envy and hate. It is sufficient for strong minorities to regress in this way for the entire social climate to change.

Republic of fear: the rise of Donald Trump

Even in the United States, the motherland of the belief in progress, the optimistic mood has turned. The ethos of freedom and responsibility, the influx of new, energetic immigrants, religious tolerance and cultural diversity, inventiveness and entrepreneurship made America great. Today, large sections of the white middle class feel threatened by these very elements. They see the opening of markets as the migration of industrial jobs to other countries, the competitive society as a danger to their social status, the mass immigration from the South as dumping competition on the labour market, the promotion of ethnic minorities as a threat to the 'White Anglo-Saxon Protestant' culture. The cosmopolitan lifestyle of Silicon Valley is worlds apart from the sad reality of the old industrial centres in the American Rust Belt and the impoverished rural regions. While 'political correctness' continues to blossom at universities, the postmodern left has lost its connection with the concerns and issues of the working class. The identity politics of the left, centring on sexual orientation and ethnicity, is now coming up against an identity politics from the right. The white heterosexual man is hitting back, and Donald Trump is his idol.

In the past twenty years, the profits of globalization have been enjoyed by a small upper class, while the income of the traditional middle classes has stagnated in real terms and private debt has grown. The main losers were the traditional industrial workers and those with low qualifications. One in five white men between the ages of thirty and fifty is unemployed. The winners are those working in the digital economy, the highly qualified and the cosmopolitan, or creative entrepreneurs and owners of capital who invest their assets worldwide. The increasing costs of children's education, medical treatment and provisions for old age, together with steeply rising costs of house ownership and rent, have meant that more and more people with average incomes find it difficult to maintain their standard of living. As in Europe, upward social mobility has declined. Class structures are re-establishing themselves and the promise of social advancement through education and work is disintegrating. This is particularly serious in a country which prides itself on the collective myth that everybody who washes dishes can become a millionaire. The American dream has become a fairy tale that is remote from the everyday reality of the large majority.

At the same time, growing sections of the population have come to believe that the costs and benefits of the liberal world order are extremely unequally divided: while the United States pays the lion's share of NATO's military expenditure and has found itself involved in extremely expensive wars in far-off world regions, the export trade deficit with China and Europe is growing. Trump's slogan 'America First' found a ready audience.

This all set the stage for the improbable rise of Donald Trump from a derided outsider to president of what is still the most powerful country in the world. It is one of the strange features of America's political culture that a New York real estate tycoon could become the mouthpiece of protest against the establishment. Unlike Germany, wealth in itself is not suspicious in the United States. On the contrary, economic success gives political credit. It makes a candidate independent of the political influence of financially powerful interest groups, one of the basic evils of American democracy. Anyone who can finance his election campaign himself is not dependent on others. In contrast to his conservative rivals, 'the Donald' struck the nerve of a considerable proportion of voters, who felt both culturally and socially marginalized. Rednecks and lowly white-collar workers form the critical mass in the protest against 'Washington'. They not only *feel* like losers of globalization, they *are* the losers.

But to explain Trump's victory over Hillary Clinton solely in terms of economic factors is not enough. Surveys have shown that his impact on supporters is based on three issues that are only indirectly connected, if at all, with economics: immigration, Islam and the 'ethnic panic' among white voters. There are deep-seated misgivings within the Republican base about the transition of America into a country that is less white and less Christian. The conservative journalist Avik Roy has concluded that 'white nationalism' is the centre of gravity of the Republican Party.[3]

Liberal intellectuals see Trump as a bizarre figure, a symbol of the political degeneration of America. But he has understood two basic things that many Democrat politicians have overlooked. He speaks to the *feelings* of an angry electorate instead of just offering them sober pragmatism. And he has an antenna for the high level of insecurity and bitterness that has built up in large sections of the population and seeks political expression. While the liberal establishment looks on the white lower class with a mixture of fear and disdain, Trump charms the little people by saying 'I love the uneducated.'

It is illustrative to look at the differences to Barack Obama's first presidential campaign. He also mobilized millions who felt ignored by the political class. But, unlike Trump, he did not rely on promises of reviving the 'good old days' but on a better future. His key message was 'Yes we can!' – together we can build a better America. But that was all such a long time ago. It reflects a strategy for countering the anti-liberal offensive. This message should not be aimed solely at an alliance of minorities. It must also include white workers and the conservative segments of the middle class.

6

The Migration Battlefield

Among the motives behind the revolt against the open society, the question of refugees and migration is the most explosive and the one that gives rise to the strongest emotions. It is a question of identity: who are we, how do we want to live, who belongs and who doesn't? The refugee question acts everywhere as a catalyst for nationalist populist movements and parties. They call for a return to tradition and the nation and at the same time fuel fears of cultural alienation and social rivalry. In both Europe and the United States, migrants have become the symbol of an uncontrolled globalization that needs to be protected against.

In his study *The Clash of Civilizations*, Samuel Huntington predicted new conflicts along old historical, cultural and religious lines. He was greatly criticized for this. Economic and geopolitical interests are also at stake when countries wage war. But looking more closely at who is fighting whom, we can clearly see the role played by the long lines of culture and religion. We should at least recognize them so as not to be at their mercy.

A persuasive argument against the understanding of cultures as collective entities is to point out the cultural struggle taking place *within* these entities. In fact, perpendicular to the supposed lines between civilizations, we find bitter conflicts between fundamentalist and modern forces. This applies to the Islamic world, and to the West and to Russia. The conflict between the open society and its enemies does not take place along geopolitical or religious lines but cuts right through the societies in question.

The conflict regarding refugees and migration is more than a struggle for scarce resources and social entitlements. It is a cultural struggle for the identity of our societies. Do we want to return to the idea of the homogeneous nation linked fatefully by the same origins and culture – or do we want to continue towards a multi-ethnic, multi-religious and multicultural society? We might not agree about what level of annual migration is manageable, about border controls and selection procedures, about the pitfalls of integration, about headscarves and the role of Islamic associations. But the basic question of whether we want an open or a closed society needs to be addressed. In reality, we have long been a society of immigrants. But that doesn't mean at all that the question has been settled.

Germany is in the throes of a demographic revolution that is changing the country in many ways. Already 20 per cent of the population are first- or second-generation migrants. In schools in the large cities, an average of a third of the children, in some cases more than half, come from migrant families. In 2015, there were more than a million legal immigrants to Germany, most from other European countries. Added to this were about a million refugees. Deducting the surprisingly large number of emigrants, the immigration balance is around 1.15 million. These are impressive figures. After Turkey, Germany is the country that has taken in the largest number of asylum seekers. Nowhere have so many requests for asylum been received – more than 440,000 in 2015 alone.

Those who prematurely declared the 'flood of refugees' to be a national emergency have been proved wrong. A decisive role is played by the extensively voluntary commitment by members of the public, who have shown practical solidarity and humanity. Many are still working today. By European standards, the acceptance of refugees in Germany is still high. But the initial optimism shifted in a few months to a sharp polarization of society. In growing segments of the population, we have seen an explosive mix of fear, prejudice and open racism, coupled with a great willingness by right-wing radical groups to use violence. According to the Bundeskriminalamt (Federal Criminal Police Office), there were no fewer than 924 violent attacks in 2015 on refugee hostels, including several cases of arson.

No isolation – no open borders

The German government has now joined a European mainstream that is concerned above all about 'securing the outer borders'. There is nothing reprehensible about that in itself. Effective control of the outer borders is a condition for open borders within the Schengen area. One is not possible without the other. It is very important, however, *how* this joint border regime is implemented: whether it aims at maximum isolation or controlled immigration. When the Austrian head of government Sebastian Kurz – a dashing national conservative – demands that 'illegal refugees' be turned away at the borders of the EU, he is reflecting mainstream thinking in Europe. Chancellor Merkel has also come round to this way of thinking. 'Illegal' in this context is anyone who attempts to reach Europe independently by land or sea. For most of these immigrants, however, there is no alternative way of getting to Europe. Whether they are recognized as refugees can only be determined in a legal asylum process.

The call for limitation became dominant in Germany when people felt that the government had isolated itself in Europe and had lost control of events. There are a number of political lessons that can be learned from this.

First, it would not be possible in the long term for Germany to go it alone in its refugee policy. Germany needs a coordinated approach with at least some, if not all, other European states. As it is unlikely that all twenty-eight members will be capable of reaching consensus in the foreseeable future (except on a highly restrictive course), the government should attempt to form coalitions with a handful of other European states that can be persuaded to pursue a liberal refugee policy based on the Geneva Refugee Convention.

Second, a minimum set of regulations and controls is required so that the public does not have the impression that immigration is completely out of control. Anyone who responds to the call for isolation with the demand for open borders for all is playing into the hands of the populists. Angela Merkel is right to refuse the fixing of binding upper limits for refugees and a return to comprehensive national border controls. At the same time, no government can ignore the need to keep immigration at a level that can be managed without great social or political disruption. Sweden provides a striking example in this regard.

The conflict between limitless hardship and limited capacities for integrating refugees is inevitable. All that can be done is to attempt to defuse it through far-sighted policies. This includes acceptance quotas for war refugees stranded in Turkey, Jordan or Lebanon.

Public discussion on a refugee treaty with Turkey continues. Is it possible to enter into a refugee pact with a president who is in the process of creating an authoritarian state and who is continuing his war with the Kurds in Syria? If not, what is the alternative?

Those wishing to reduce the uncontrolled flow of refugees to Europe must expand the legal access corridors. Both of these aims can be achieved only in cooperation with states neighbouring the EU that have absorbed most of the refugees from Syria and Iraq or function as a transit bridge to the EU. The criterion is not whether we like these regimes but whether the agreements are reconcilable with the Geneva Refugee Convention and European law. It is a question of improving the situation of refugees where they are and at the same time of the legal entry into the EU of an appreciable number of stranded persons. If this package cannot be implemented, we will sooner or later once again see thousands of refugees in overloaded boats on the way to Greece and Italy.

Third, those who wish to uphold the right to asylum must also use it properly. This calls for a rapid distinction between people with a reasonable claim to asylum and refugees for social reasons. War refugees are also not migrants as such. A decision must be made as quickly as possible as to who should be offered only temporary asylum and who has a prospect of long-term residence. For this to happen, the change from refugee to migrant status must be made easier. It is legitimate for people to try to escape hardship in their homelands and to seek a better life in Europe. But it asks too much of the asylum law for it to be used as a loophole for all conceivable migration motives. The possibilities for regular labour migration must therefore be expanded – not only for highly qualified persons but for everyone with the prospect of a job in Germany. Another possible solution would be the issuance of rotating work permits for a limited period (commuter migration). If labour migrants return to the countries of origin with new professional qualifications, this lessens the negative impact of the exodus of young dynamic people from poor countries.

Combating the reasons for flight instead of the refugees

According to the United Nations High Commissioner for Refugees, more than 65 million people are currently in flight – more than at any time since the Second World War. Two-thirds of them are internally displaced persons. This humanitarian drama cannot be solved by opening the way for all refugees to Europe. A sustainable refugee policy must start by combating the causes. No one should be forced to leave their homeland. Admittedly, this is easier said than done. In spite of its economic, political and military resources, Europe does not have the power to end all wars and all misery in the world. Claims of this sort are left-wing colonial thinking, a kind of moral delusion of grandeur.

Even if the remote effects of our consumption are felt around the world, from the palm oil plantations in Asia to the cobalt mines in the Congo, we are no longer dealing with mere objects of colonial rule. It is the ruling elites in the South that are responsible for the wealth and misery in their countries. Without their assistance, no international business can purchase land, operate mines or clear forests. In most countries of the global South, the gap between the very rich and the very poor is wider than in the old industrial countries. Corrupt elites amass fabulous wealth, which they send to Switzerland, Panama or London, instead of investing in their own countries. In the last few decades, the flight of capital from Africa has been greater than foreign development aid. Thabo Mbeki, the former president of South Africa, puts the annual export of capital from Africa at 50 billion dollars – money that is urgently needed on a continent faced with mass poverty, miserable infrastructures, poor schools and decrepit hospitals.[1] Around two million people alone have fled Zimbabwe, which the former freedom fighter Robert Mugabe has driven to the brink of ruin. There are also refugees from countries like Nigeria and Angola that are rich in resources, where income from oil production disappears into the pockets of international businesses and a small local upper class, while large parts of the population are living on or below the poverty threshold. Nigeria is one of the main sources of capital flight from Africa.

The main instrument for improving the social situation of the countries of the South is no longer development aid but better governance: the rule of law, combating corruption, transparency of financial flows, government accountability, independent media and free elections. Only when the ruling elites come under democratic pressure from their subjects will they invest more in education, health, public infrastructure and agricultural development. The EU can promote this change by linking its development and trade policies more with the reform of government and the administration.

The European foreign trade policy is also in need of critical revision. It is cynical to build fences to keep out refugees while we destroy the livelihood of small African farmers with export subsidies for European agricultural surpluses. Binding sustainability criteria should be fixed for the import of natural resources and agricultural products so as to put an end to the ruthless exploitation of people and nature. Agreements of this type on the certification of wood, food, agricultural resources, textiles and diamonds have already been concluded between non-governmental organizations and international enterprises. They could serve as blueprints for binding regulations by the EU. Stricter transparency rules for payments in the natural resources sector would make an important contribution to combating corruption in the system. This will not happen overnight, and it is not at all clear how much such measures would reduce the number of migrants. More effective control of the EU outer borders, rapid conclusion of asylum procedures according to uniform European standards and return agreements with countries of origin are essential. But combating the causes of flight is certainly a more humane way of dealing with refugees.

Immigration and integration

In its own interests as well, Germany must develop a far-sighted immigration policy. Immigration is not a patent solution for demographic change. But without a considerable amount of immigration, the labour potential is likely to shrink drastically in the next decade, while the number of pensioners is growing at the same time, increasing the cost of health and care. A massive shift of this nature in the social balance could well lead to a loss of prosperity and rancorous distribution conflicts. These effects can be at least attenuated by a long-term immigration policy. The immigration of young, motivated people can only be of benefit to the economy of an ageing society, providing, of course, that the new arrivals are successfully integrated into the labour market.

Is it legitimate to attach such importance to immigration? Altruistic arguments will not be enough to win a political majority. A liberal migration policy will be supported in the long term by the population only if it is seen to offer advantages for both sides. The refugee question is different. Here we must insist that people fleeing from war and persecution have a right to protection, regardless of any utilitarian motives. We should not make cost-benefit calculations in the case of refugees. This is not to say that refugees could not also be a bonus for our country if they are integrated quickly enough into the labour market. Expenditure on professional training for refugees is an investment for the future.

Immigration legislation is overdue for both symbolic and practical reasons. The next logical step would be to change the German citizenship law from *ius sanguinis* to *ius soli*. Being German would then no longer be a question of ancestry: German citizenship could be acquired by being born in the country or by obtaining it as an immigrant.

Going hand in hand with a liberal immigration policy is a long-term integration policy. However controversial the term is, it would be senseless to abandon it. Migrants are not arriving in a no man's land but in a society with a specific history, language and political culture. Those who are not willing to embrace it would be better off looking elsewhere. Successful integration means transforming foreigners into citizens. This demands a willingness by immigrants to accept the country, its history and its constitution. At the same time, it requires that the existing population accepts the new arrivals as equals. No one should be forced to disavow their origins, language or religion. Discrimination and racism cannot be tolerated. They violate human dignity and contaminate life together. Integration is not the same as adaptation. Successful integration calls for changes on all sides. In this way of thinking, it is all the more evident that certain things are non-negotiable: the basic values of the constitution and the liberal lifestyle in Germany.

We should not confuse a welcoming culture with welfare paternalism. Immigration societies work when immigrants have the opportunity of advancing under their own steam and of creating a better life for themselves and their children. The key to this is education, work and political participation. By contrast, the endless debates about headscarves or dual citizenship are just shows of force. It is not a sign of self-assurance when an old German demands of newcomers that they cut their ties with their countries of origin. 'Observe our laws, learn German, look after your children's education, be good neighbours' is quite sufficient. Otherwise everyone has the right to be happy in their own way, as long as they do not infringe on the freedom of others.

7

Dealing with Islam

Those who wish to convince others must be sure of their own values and traditions – not in a bigoted way but with the calm conviction that they have a better answer to the challenges of our time. The democratic awareness of the Germans is not in good shape, however. The attitude to politics fluctuates between self-flagellation and arrogance, between pride in German failings (no one was worse than the Germans were) and self-congratulation (no one is more moral than the Germans are). Sober criticism aimed at a pragmatic improvement in the situation is not in great demand, but passionate appeals for radical and immediate change are all the commoner as a result: we should immediately change our thinking, our way of life, our relationship with the rest of the world – not to mention capitalism. The old German mixture of idealism and radicalism is seen as a lofty ideal, while pragmatism and compromise are despised.

The urge to self-flagellate is to be found in educated circles, which still see the dark powers of patriarchy and racism in every aspect of German life; in repentance preachers in green mantles, who regard a modest and self-sufficient life as the only way of avoiding a climate disaster; and in angry old men, who attribute all the evils of the world to the wickedness of the West. Even the terrorist attacks by Islamic jihadists are seen as a reflection of the crimes committed by the rich North against the poor South. It is our own fault if the victims of western imperialism declare war. The reverse side of this embittered self-criticism is Eurocentrism, a distorted belief in the omnipotence of the West. There is always this duality between the perpetrators (the white man, imperialism) and the victims (everyone else). It can be overcome only if the West is pushed off its arrogated throne. In that regard, the main accusers of the West at home agree with Islamic fundamentalists, anti-western Russians and radical nationalists in the former colonies.

The reverse of this self-flagellation, in which all progress in civilization disappears in a single disaster scenario, is the reluctance to defend values when they are challenged. The sharpness of this self-criticism is in striking contrast to the understanding that large sections of the public show for despotic regimes and anti-western ideologies. Seldom is the criticism of Putin's Russia or the mullah dictatorship in Iran as sharp as it is of Trump. When a couple of Mohammed cartoons in a Danish newspaper lead to violent demonstrations and wild threats from the Islamic world, we discuss whether the newspapers were an irresponsible provocation. Fundamental criticism of the West has a ready audience. Is there not some truth in the Islamic criticism of western materialism, liberal whateverism, the replacement of spiritualism by consumption, the renunciation of family values and the sexualization of public life? Why does the United States always have to interfere in areas that are not its concern? And is the miniskirt really an expression of female emancipation or just a symbol of the transformation of women into sex objects? Germany must finally stop imposing its values on the rest of the world. This sounds like a parody, but it has long become part of daily discourse.

Houellebecq: *Submission*

If literature is a seismograph for the undercurrents in society, Michael Houllebecq's *Submission* immediately springs to mind. By a macabre coincidence, the novel appeared on the day of the attack on the satirical magazine *Charlie Hebdo* and the Jewish supermarket in Paris in January 2015, which made the book appear as a clairvoyant anticipation of the unstoppable Islamization of France. Critics were correspondingly charged – some accused Houellebecq of encouraging Islamophobia, others read the book as a satire on the fear fantasies of old Europe of being overwhelmed by an 'Islamic invasion'.

For me, *Submission* fits into neither of these categories. The curious aspect of the story is that both the Islamists and their opponents, the 'identarian movement', converge in their rejection of liberal democratic capitalism. Both see it as the cause of the crisis in Europe. The democratic republic is now just a caricature of itself, hollowed out and exhausted, its elites powerless and opportunistic. It is ripe for a bloodless enemy takeover.

The central intellectual figure in the novel, the recently appointed president of the Islamic University of the Sorbonne and later foreign minister of the Islamic Republic of France, converts from being a mastermind of the populist nationalist movement to the intellectual figurehead of the Muslim Brotherhood. For him, 'female' Christianity has lost its power for renewal, which he sees in 'male' Islam. The Christian churches have forfeited their spiritual power; they indulge in abstract humanism and offer the framework for a completely secularized culture. Only Islam, which does not accept the separation of church and state, still has the power and will to offer an alternative society. The flaccid Europeans should accept that their last chance for a great future lies in a 'Eurabic Union' with the newly awakened Muslim world.

That neither the conservatives nor the left in Houellebecq's novel has any answer to political Islam is not so far-fetched. Many conservatives share with it their discomfort at the secularized modern world and the disappearance of the traditional gender order, many leftists its anti-Americanism and sympathy for a state-run economy purged of the sovereignty of money. Both are susceptible to an imagined community in which the individual is subordinate to a higher order. To that extent, Houellebecq describes political Islam *not* as something alien conquering Europe 'from without' but as a concomitant to the anti-liberal movements within Europe itself. The democratic republic does not need to be conquered by force; it is all too willing to give itself up.

The longing for a fixed order, stable social relations and an 'organic social order' is the foundation on which the submission of the republic under Islam rests. The analogy to the collaboration with the Nazis during the Second World War is all too clear.

The novel culminates in the choice between two anti-liberal movements: the Front National and the supposedly moderate Islam. The scales are tipped by the spavined socialists and conservatives, who enter into an 'extended republican alliance' with the Islamists. Once they have won their victory, France turns practically overnight into an Islamic republic: short skirts disappear from the streets, women are sacked from public institutions, the education system is confessionalized, the major universities are taken over by

Islamic foundations. Everyone has their place in this society – except for Jews and homosexuals. Local family businesses become the backbone of the Islamic economy. The old left–right order within the political landscape collapses, leaving an opposition composed of a handful of left-wing secularists on the one hand and radical Salifists on the other.

At its core, Houellebecq's novel is an angry parable on the crisis of liberal democracy. Socialism and conservatism are both just empty shells, the old political elites are tired, and public debate has become a routine ritual. The economy is stagnating, while the elites tenaciously defend their privileges. The victory of individualism has created a society of monads incapable of connection. There are no great narratives any more to generate enthusiasm and political solidarity. This vacuum is being filled by identarian movements of different hues.

In passing, this book offers a sober look at the opportunism of educated elites. The life of the protagonist, a literature professor undergoing a midlife crisis, is one of academic routine, erotic escapades, self-pity, hypochondria, half-hearted attempts at self-conversion to Catholicism and a final submission to the new regime. He ultimately follows the call to an Islamized Sorbonne financed by Saudi Arabia and is willing to convert to Islam for that purpose. His motives could not be more profane: a princely salary and the prospect of at least three wives, in other words, the promise of sex, without the nuisance of courting, and regular well-cooked meals at home. In the end, the Islamization of Europe turns out to be a banal male fantasy. Its impact is unfortunately not confined to literature. ISIS would scarcely be so attractive to Muslims in Europe if membership of this warrior order did not also offer a licence for the sexual exploitation of women shared out as the spoils of war.

Houellebecq does not demonize Islam. His criticism is of the cynicism of the ruling elites and the powerlessness of the democratic republic. In place of this, political Islam is presented as a youthful and persuasive movement. It impassions an international, wide-ranging brotherhood dedicated to the victory of Islam. Unlike the flaccid liberals, Islamism subscribes to the will to power. It is no coincidence that the newly converted model intellectual in the integrationist movement is a disciple of Nietzsche.

Pre-emptive obedience

What rings a bell in *Submission* is the strange tiredness of European democracy, the absence of political passion and the willingness bordering on self-sacrifice to relativize one's own values. The opponents of liberal democracy are full of combative spirit, fighting their cause with energy and not baulking at conflict, while the elected representatives of democracy do their utmost, on the pretext of diplomatic wisdom, to avoid any conflict.

When the Iranian president Hassan Rohani came on a state visit to Rome in January 2016, he had contracts worth billions for Italian companies in his luggage. This is the best guarantee for a warm welcome. Between the meeting with the Italian prime minister and the subsequent press conference, the protocol provided for a visit to the Capitoline Museums. During the visit, Rohani was kept away by zealous Italian authorities (it is unclear which one was actually responsible) from the famous Capitoline Venus. So as not to embarrass the distinguished guest from the Islamic republic, the statue was simply boarded up. It made no difference that she covered her private parts with her hand. Many other statues presenting themselves in their classical nakedness suffered the same fate. It may be supposed that the Italian officials wanted to impress Rohani, as a representative of a millennial power, with their own cultural tradition, from empire to empire, so to speak. At the same time, however, they were all too willing to deny this tradition. It was not the first faux pas of this type. A little earlier, the head of government Matteo Renzi had received a prince from the United Arab Emirates in Florence, who had rescued the Italian airline Alitalia from bankruptcy. Before he was shown the Palazzo Vecchio, a naked faun, a contemporary work by the star artist Jeff Koons, was hidden behind a screen. These bungling efforts would be laughable if they were not symptomatic of an attitude of mind that confuses courtesy with submission and considers pre-emptive obedience to be a sign of intercultural competence.

The Mohammed caricature controversy

The confrontation with the laws and prohibitions of fundamentalist Islam might have been farcical in Italy, but in Denmark and France it was deadly serious. In autumn 2005, violent demonstrations took place in several Islamic countries against the publication of a series of Mohammed caricatures in the Danish newspaper *Jyllands-Posten*. One of the drawings showed the Prophet with a bomb flaunting the Islamic confession of faith. The conclusion was obvious: Islam equals terror. In view of the real diversity of Islam, this is obviously nonsense. The question here, however, was not whether this exaggeration was accurate but whether it was *admissible*. If the publication of blasphemous caricatures is covered by the freedom of art and of the press, it must also be tolerated in the case of Islam, as the countless drawings and pamphlets that have hurt the feelings of devout Christians have always been.

Like every freedom, the freedom to mock is not limitless. Many democracies have blasphemy laws to protect religion. In Denmark, those who 'publicly mock or deride the doctrine or worship of a religious community existing legally in this country' are subject to a fine or imprisonment. Whether such protective rules serve public order or should be cancelled as relics of the alliance between the throne and the church is a matter of debate. In case of doubt, the balance between art and press freedom and the protection of religious convictions should be decided in court. At all events, it is no justification for murder and manslaughter.

When the caricature controversy first broke, there were plenty who criticized the editor-in-chief of *Jyllands-Posten* for provoking Muslim sensitivities. Only a handful of western media went as far as reprinting the caricatures in a demonstrative defence of freedom of the press. There is a thin line between the appeal not to fan the flames of Islamism and pre-emptive self-censorship by the West. Explained as thoughtful consideration for the sensitive pride of Muslims, it also has its origins in the naked fear of the threats of violence by radical Islamists. In this way, the danger of Islamophobia is transformed directly into a willingness to allow Islam and its representatives special rights.

This attitude has nothing to do with respect and equality. On the contrary, Muslims are treated not as responsible people capable of responding to polemic criticism in a civilized way but as emotionally charged time bombs that should not be provoked at any cost. Those who say that texts and caricatures critical of Islam should not be published out of consideration for the feelings of Muslims are basically agreeing with the thesis that Islam and democracy are irreconcilable. The attempt at appeasement in the face of the aggressive intolerance of parts of the Muslim community encourages the intolerant. This also applies to attitudes to Christian fundamentalists and those who advocate limiting freedom of expression by referring to divine commandments or to respect for the feelings of others. The question is whether Islam and its radical versions can be subject to the same biting criticism as has been directed for centuries at Christianity, the institutional church and its representatives.

In contrast to the conflict surrounding the Danish Mohammed caricatures, the murderous attack on the editorial board of the French satirical magazine *Charlie Hebdo* in January 2015, in which twelve people were killed and twenty others wounded, provoked a wave of solidarity. This might have had something to do with the brutality of the attack, the popularity of the illustrators and the place where it happened. The murder of journalists in the centre of Paris who fail to show respect for the Prophet and his supporters hits at the heart of democracy. It is encouraging that millions of people in France and many other countries saw this attack as an assault on European values. The following issue of the magazine, which appeared with the title *Le Journal des Survivants* (the magazine of the survivors), sold seven million copies and was translated into sixteen languages. The front page showed a cartoon of Mohammed with a tear in his eyes holding a sign saying 'Je suis Charlie' and the caption 'Tout est pardonné' (all is forgiven).

Once again, however, there were those who said that the victims also shared at least some responsibility for their fate. Henri Roussel, co-founder of *Charlie Hebdo*, said that the slain editor-in-chief Stéphane Charbonnier had his colleagues on his conscience. He asked 'what made him feel the need to persuade the team to go to such an extreme.' He had previously accused 'Charb' of having transformed the magazine 'into a Zionist [sic] and Islamophobic organ'. Nicolas Blancho, president of the Islamic Central Council of Switzerland, accused the journalists in the first issue after the attacks of intellectual incendiarism that was 'just as dangerous as extremism'. The then Turkish prime minister Ahmet Davutoğlu condemned the front page of the *Journal des Survivants* as a 'serious provocation' and repeated the accusation of 'insulting the Prophet' that was tantamount to a licence to embark on new punitive actions. Pope Francis warned that the beliefs of others should not be made fun of. Bernd Matthies, long-serving columnist in the Berlin *Tagesspiegel*, commented that the terrorist attack would 'very probably not have happened if Charbonnier had decided to focus his satirical attacks more on the French government, on Marine Le Pen or other key figures in politics.'[1] In other words, it is acceptable to make satirical attacks on one's own government or 'against the right', but not against violent supporters of the fundamental teachings of the Prophet. Those who anger Islamists should not complain when they retaliate. If anything, it is the victims' own fault if they are murdered by fanatic defenders of the Prophet. Those who argue in this way have already abandoned the notion of freedom of the press.

What does this have to do with Islam?

The cautious restraint in addressing Islam and its supporters is also shown in the wide berth given by many politicians and media to critical Muslims, particularly those who dare to break the taboo and question the idea that violent terrorism, intolerance and the *déclassement* of women have 'nothing to do with Islam'. The same arguments could be used to claim that the medieval crusades, the Inquisition, European anti-Semitism or the outlawing of homosexuals for centuries by the Church had nothing to do with Christianity.

The worry that critical discussion of the aberrations in Islam could foster the latent hostility to Muslims is understandable. But the opposite result is achieved by making the public believe that there is a clear distinction between the true and peaceable Islam and all the terrible acts carried out in its name. The public notes the intention and is disgruntled. Those who put a taboo on sensitive questions so as to protect minorities pave the way for populist taboo-breakers. This also applies to the discussion on Islam as a political religion. How much Islam there is in Islamic State, in the Taliban, in the Islamic Republic of Iran or in Hamas is a question that should be discussed openly. It would be just as ill advised to take their appeal to 'true Islam' at its face value as it would be to claim that it is a mere subterfuge.

It is misleading to attribute certain cultural practices and political phenomena to 'Islam' or 'Christianity' – if only because there is no monolithic Islam or Christianity. It is fairly obvious that religions change with the times. Their holy scriptures can be interpreted in different ways and offer a treasure trove for opposing political directions. It should be left to religious scholars to discuss whether violence towards non-believers, the marriage of young girls, or the whipping of homosexuals derive from Islamic traditions, whether Mohammed preached the submission of women or was perhaps a covert feminist, whether Islam is inherently warlike and to what extent the separation of state and religion is reconcilable with Islamic teaching. The answers to these questions are likely to be just as diverse as the interpretations of Christianity. There are passages in the Bible and the Koran to demonstrate whatever you are looking for – be it doctrines of compassion or revenge, love or intolerance of non-believers, peace or war. It is not for parties and politicians to make judgements on religion. Their task is to reject any general suspicion of Muslims, to protect their freedom of religion and at the same time to defend democratic norms against any religious pretensions.

Freedom of religion and human rights

No religion is above human rights and outside the law. Freedom of religion is a fundamental right. This does not release the democratic state from its obligation to set legal limits to prevent other fundamental rights from being violated in the name of religion. There are good reasons for tolerating the wearing of headscarves by Muslim teachers as an example of freedom of religion in practice. A ban on headscarves in public institutions would place devout Muslim women in an existential conflict between religious faith and career. In many cases, it would be tantamount to an employment ban. Instead of promoting the equality of Muslim women, a headscarf ban slows down their emancipation.

We should not be taken in by the argument, however, that the difference between the hijab and the burqa is just one of degree. To hide women completely is a violation of their human dignity and the ban on discrimination in the Basic Law. It is absurd to make tolerance of the burqa in public institutions into a question of religious freedom. Total concealment is an instrument of power by men over women, a symbol of their relegation to second-class beings. The fact that there are Muslim women who see their degradation as an act of free choice does not make things any better. There are women who defend the act of female genital mutilation, but it does not stop it being a brutal form of bodily injury.

It is irrelevant to say that the fight against misogynistic customs and traditions cannot be won primarily through prohibitions. It is not the purpose of laws to convince perpetrators but to protect victims. The often-used argument that it only concerns a small number of women is also irrelevant. Since when has the assertion of human rights been a question of numbers? There is a strange reversal in the discourse on women's rights as soon as Islam is mentioned. No quarter is given in discussions with the Catholic Church, with conservative family politicians and Christian anti-abortionists. The control by women of their lives, their bodies and their sexuality is an immutable value, not just for feminists. But as soon as the Muslim minority is questioned, some members of the liberal public do an about-face. Then the focus is no longer on the inalienable rights of women, but on the worry that it could encourage anti-Islamists if certain cultural practices were said to be unacceptable.

With this logic, the banning of the burqa and niqab from public institutions is transformed from a measure against discrimination against women into an act of discrimination against Muslims. Its moral legitimation derives not only from its appeal to absolute freedom of religion, claimed in this form only for Muslims, but also from its anti-colonial aspect. Islam is the religion of immigrants from the former colonies. They are seen as the victims of the white man, who need to be protected against continued racism.

It is for that reason as well that the post-colonial intellectual milieu had such difficulty taking a clear line on the attacks by Muslim men in Cologne on New Year's Eve: its concern was less with the actual victims of sexual violence than with the feared stigmatization of North African immigrants. This fear is not unfounded – racism and Islamophobia are real phenomena. The mistake is to apply different standards to migrants than to locals on the pretext of anti-racism. It is misleading to ignore the specifically misogynistic culture in conservative Islam societies by referring to the daily sexism in western societies. Those who sugar-coat the real problems connected with the immigration of men from patriarchal Islamic societies only feed the resentments they are trying to combat.

Islam is extensively unknown territory for most non-Muslims. At the same time, radical offshoots of this world religion are violently breaking into our world. It is therefore all the more important to discuss in public why Islamic extremism is so attractive to thousands of young people and how we can confront it in a calm but determined manner.

It remains to be seen whether the idea of a specific 'Euro-Islam' that assimilates the basic European democratic values is more than just wishful thinking. The many modern Muslims in Europe demonstrate that it is not far-fetched. Sadiq Khan, the directly elected mayor of London, could act as a symbol for an Islam that is receptive to modern liberalism. He is not an isolated case. Before the murderous violence against Muslim Bosnians in the Balkan wars, Sarajevo was a tolerant multicultural city. Photos of Kabul or Tehran in the 1960s show self-confident young women with short skirts and jaunty hairstyles. They look today as if they belong to another era. The crisis of secular Arab nationalism and the failed authoritarian modernization have led to a radicalization under the banner of Islam whose repercussions are also being felt in Europe. It no longer aims at catch-up modern reforms but at a militant demarcation from the West.

The reform of Islam must take place from within. This is conditional on a climate of freedom of thought in which criticism and self-criticism are not persecuted as heresy or dismissed as Islamophobia. Those who support the integration of Islam in Germany must encourage public debate among and with Muslims. If Muslims are capable of what we ask of everyone else – critical discussion of their religion – then they will certainly be welcome as citizens with the same rights and obligations.

The warning from Syria

At the same time, the West must reorientate its policy towards the Middle East. The attitudes, ranging from indifference and chumminess with despots to military interventions against them, have given rise to mutual alienation. No one knows any longer what to expect of the West. The extreme brutality of the war in Syria has put the finishing touch to any respect for Europe and the United States in the region. If the intervention to overthrow the Saddam regime in Iraq was an error bordering on a crime, the passivity in the face of the continued massacre by Bashar al-Assad of his own people is no less of one. The opportunity for the United States, France and the United Kingdom to impose a flying ban on the Syrian air force and to establish safe withdrawal areas for the population has been missed. As a result, the chance for an orderly transition of power has been wasted and the door opened wide to Russian intervention. The West's indifference has driven moderate opponents of Assad into the arms of radical Islamists.

Hundreds of thousands of dead civilians, millions of displaced persons, the devastation of entire cities – all of this suffering and hate also come back to us. It is quite possible that our indifferent attitude to Syria could influence the image of the West in the Arab world for some time to come. After the first wave of empathy for Syrian refugees in Sweden and Germany, the willingness to accept them is rapidly dwindling. The European Union is cutting itself off. But even a more open attitude to refugees cannot replace an active Middle East policy. In reality, the West has made itself dependent in Syria on Putin. It looks as if ultimately an arrangement with the butcher Assad will be sold as a peace settlement. This would not only be a further blow to the West's credibility as the custodian of human rights. It would also be bad realpolitik because long-term stability in Syria cannot be obtained with the Assad regime.

A Marshall Plan for the Middle East

The West will have to accept the difficult balancing act of cooperation with authoritarian regimes in the region and a policy based on human rights. Diplomatic arrangements with the existing autocracies are necessary, as with Putin's Russia. These include security agreements, trade agreements, scientific and cultural exchanges. They must be designed to promote a long-term process of modernization. The key to the region's future are the millions of young persons. They are better educated than their parents but have wretched career prospects. The number of young Arabs between the ages of fifteen and twenty-four is set to rise from the 46 million in 2010 to 58 million in 2025. Youth unemployment is already worrying; in Egypt it is put at 40 per cent. As long as a majority of university graduates have no other prospect than a life in poverty or emigration to countries where they are not welcome, radical Islamists will have a field day. At least they offer the promise of identity, a sense of purpose and pride.

The EU must put pressure on the ruling regimes to start the overdue reform of the inflated public administration and an economy hampered by state bureaucracy. There is no chance of prosperity and stability without progress towards the rule of law and a market economy. In return, massive investments should be offered to modernize the education system, energy sector and infrastructure. The demand for a 'Marshall Plan' for the Middle East might sound far-fetched, but it is precisely what is urgently needed: a programme for modernization of the economy, hand in hand with the development of rule of law and democratic structures. This policy is in Europe's very own interests.

The attempt to suppress the crisis in the Arab world with unyielding violence is doomed to failure. If the West tethers itself again in the name of stability to despotic rulers like Egypt's President Sisi or Syria's Assad, it will forfeit the last remnants of political credibility. This also applies to relations with the ruling powers in Tehran. There is a fine line between necessary cooperation and unprincipled chumminess. If the West associates with regimes that trample on human rights and rule by violence both within and without, it is giving up on itself.

8

No Empathy for Freedom: The Germans and Ukraine

There is a widely held attitude in Germany of 'serves you right' when it comes to the Russian intervention in Ukraine. From the earliest days of the Maidan movement in November 2013, when a growing number of people protested in favour of an association agreement with the EU and against the corrupt Yanukovych regime, there were those who were already warning of a conflict with Russia. The Ukrainians should accept that they were inevitably part of the front yard of their mighty neighbour. Russia would never allow Ukraine to attach itself to the West. And besides, Ukraine itself was half Russian, and any attempt at European-style democratization would lead to civil war. According to the Russian propaganda booming out of the loudspeakers and its echo chambers on the internet, the revolution in Kyiv was not a democratic one; in reality it was about the restoral of the 'Bandera fascists', who once allied with Hitler's army against the Soviet Union. The Maidan was full of anti-Semites and reactionary nationalists. The overthrow of Viktor Yanukovych was an unconstitutional putsch instrumentalized by the United States.

From the left to the national conservative parties, there is general agreement that the Ukraine is itself to blame for the loss of Crimea and the war in Donbas. If it had accepted its historical fate as a Russian protectorate, the European world would still be in order. This is also the view of the Committee on Eastern European Economic Relations, whose chairman cannot bow low enough when Putin grants him an audience.

The ambivalent and sometimes hostile attitude of the German public to the Ukrainian revolution is one of the disturbing experiences of the past few years. Why is there this lack of sympathy for a freedom movement? What makes so many rational people pounce so eagerly on all the unpleasant side effects and close their eyes to the major demand of the Maidan for the creation of a democratic European Ukraine? How is it that the importance of marginal ultra-right groups is hugely inflated, although they have much less support in elections than extreme right-wing parties in Western Europe? And what, by contrast, prompts the great understanding for the undeclared war that a mighty Russia is waging against its insubordinate neighbour to force the country back into the Russian sphere of influence? Why does next to no one remember that in the Budapest Protocol during the nuclear disarmament talks in 1994 Russia, together with the United States and the United Kingdom, guaranteed the inviolability of Ukraine's borders? While every violation of international law by the United States is vigorously condemned, there is a great willingness in the case of Russia to place the right of might over the rule of law. The annexation of Crimea is not officially recognized, but people are prepared to accept it and the establishment of a Russian protectorate in Donbas, if only a cosmetic pseudo-solution can be found.

A lack of sympathy for the freedom struggles in the east of Europe is not new in Germany. Free spirits like Heinrich Heine might have spoken in support of

Poland's fight for freedom, but the great majority were quite happy to see Poland carved up by Prussia, Austria and Russia. The uprising in East Germany on 17 June 1953, the Hungarian Uprising in 1956 and the Prague Spring in 1968 were followed in the old West Germany with sympathetic interest – with the exception of the pro-Soviet left. But when Polish workers occupied Gdańsk shipyard in 1989 and a widespread protest movement rose up in Poland, the reaction in Germany was very restrained. It was the high point of the new German Ostpolitik. Having originated as a strategy to overcome the division of Europe ('change through rapprochement'), the emphasis now was on stabilizing the status quo. Large sections of the political class did not want to jeopardize the painstakingly developed relations with the communist leaders in Eastern Europe. Workers' protests and democratic rebellion were not part of the Ostpolitik script. They unsettled the balance in relations with Moscow – as do the attempts today by Ukraine to escape from the Russian cosmos.

Bothersome Ukraine

The perception of Ukraine as a nuisance factor in relations with Russia is a key element when looking for an explanation for the coldness towards the Ukrainian independence movement. Added to this is the very short memory regarding Germany's blame and responsibility for the crimes committed by the SS and Wehrmacht against the Soviet Union during the war. Any collective feeling of guilt by the Germans is generally in connection with Russia but not with Ukraine. Barely any mention is made of the fact that more people were shot, gassed, starved to death and deported as forced labourers in the 'blood lands' of eastern Poland, Byelorussia and Ukraine than anywhere else. On the territory of present-day

Ukraine alone, around 3.5 million civilians fell victim to the Stalinist terror between 1933 and 1938. They were shot or deliberately starved to death. A further 3.5 million were slain by the Germans between 1941 and 1944. And on top of that are 3 million Ukrainians who died as soldiers in the Red Army, or as partisans or civilians in the Second World War.[1] Despite all this, Ukrainians are usually seen in left-wing discourse as collaborators, Hitler's helpers, slaughterers of the Jews and concentration camp hangmen who have only themselves to blame and are not worthy of solidarity. The fact that millions of Ukrainians suffered under the Germans and fought against them does not really fit the picture and is not worth taking into consideration.

A third explanation can be found in the nature of the Maidan revolution itself. For many Germans, it is alien and suspicious because it does not match their idea of a progressive movement. It is not anti-capitalist but aimed on the contrary at the establishment of a modern market economy, at national self-determination, and it was directed against Russia, not the United States. The right to independence and sovereignty is approved of to correspond with their worldview.

Not even the sea of European flags on the Maidan has been able to influence public opinion in Germany. Even today, the vast majority of people in Kyiv, Lviv and Kharkiv see Ukraine's future in the European Community. It is no exaggeration to say that thousands of Ukrainians have given their lives for the European ideal, while within the EU enthusiasm for the united Europe project is waning. People do not want any further expansion of the EU, additional encumbrances for the sake of other countries, and above all they don't want problems with Moscow. The fact that Ukraine could be a great gain for Europe doesn't count. And yet the country has not only enormous economic potential, but it could also be a reliable base for European interests around the Black Sea. Even more, it plays a key role in the future development of Russia. A democratic and economically successful Ukraine with close ties to the EU would have an attraction for its Russian neighbours that cannot be overemphasized. It is for that very reason that Putin is attempting with all his might to prevent the success of a 'Euromaidan'. Those who have not abandoned hope of a European Russia should do everything to support Ukraine on its way to Europe.

9

The Russian Complex

Anyone who studies Germany's relationship with Russia will find out a lot about the Germans and their ambivalent relationship to freedom. Germany and Russia is a huge narrative. In spite of two wars leading to the verge of annihilation, in spite of four decades of Cold War between the Soviet Union and the West, there is a mutual attraction that has survived all hostility and all atrocities and that might even have become stronger as a result of them.

Among all the nations that fought against Nazi Germany in the Second World War, Russia has a special place in the German culture of remembrance. In the collective tradition of the war generation, the Americans attacked the German Reich without good cause. The British fought for the survival of their empire. They are both still resented for the bombing of German cities. The Nuremberg Trials, supposedly managed essentially by the Americans, were long regarded as the victors' justice. The attitude of the victorious western powers was characterized essentially by the demand for reparation and exoneration. In East Germany, 'imperialism' (basically the United States) was in any case the official enemy. The tenor was anti-fascist, but the government built on the enemy stereotypes of the Nazis, the 'Anglo-Saxon plutocracy'. No one should be under the illusion that these attitudes have simply disappeared. They continue to make trouble under the facade of political correctness.

When Germans and Russians meet, they are fond of sentimentality. When the prima ballerina dances, the violin sings and the vodka flows, everything is forgiven and forgotten. Then they drink to 'German–Russian spiritual kinship', the unholy concord that contrasts romantic introspection with western rationalism, sensitivity with calculation, tragedy with hedonism and true culture with the materialistic civilization of the West.

There is a deep-seated impulse in German–Russian politics to create a German–Russian *special relationship* independent of Russia's internal constitution and realpolitik. Russia can be what it wants, its leadership can do what it will, but German policy seeks close economic cooperation and political partnership. It is true that there are also calls for a special relationship with Israel, once again with reference to our historical responsibility. But there is no widely held sentimental fascination with Israel in Germany, nor is the attitude to the Jewish state unaffected by the policies of the Israeli government. On the contrary, the German media (and even more so friends meeting regularly over a glass of beer) are uncompromising when it comes to criticizing Israel's real or alleged misdeeds.

History as politics

How is this anomaly in Germany's view of Russia to be explained? An important role is played by the continued feelings of guilt by many Germans regarding the campaign of destruction waged by the Wehrmacht and SS against the 'Slav Untermensch'. Anyone visiting the memorial to the German siege of Leningrad, in which almost a million people starved or died in the artillery bombardment, cannot help but feel a continuing responsibility to the descendants of the victims. But the German culture of commemoration is skewed: the war of extermination against the Soviet Union is reduced to a German–Russian tragedy. All of the sympathy is directed at Russia. The Byelorussian, Ukrainian and Caucasian victims of the war are mentioned in passing, if at all.

In Russia's neighbouring states, people have not forgotten that Hitler and Stalin came to an agreement in August 1939 on the division of Central and Eastern Europe. The Molotov–Ribbentrop Pact was the starting signal for the Second World War. On 1 September, the Wehrmacht invaded Poland, and on 17 September the Red Army crossed the Russian–Polish border. The two armies took part in a joint victory parade in Brest-Litovsk. Until the Germans attacked the Soviet Union in July 1941, hundreds of thousands of Poles were deported from the Soviet-occupied territories, tens of thousands of officers and members of the political elites were shot. Similar occurrences took place in the Baltic States. To ignore this prelude to the Great Patriotic War is to ignore the historical memory of the Central and Eastern European nations caught between the millstones of fascism and Stalinism.

That entente between Germany and Russia as a guarantee of peace and security in Europe is a recurrent refrain in German–Russian dialogue. Among our eastern neighbours, however, the mention of the 'Berlin–Moscow axis' is more likely to set off alarm bells. The division of Poland between Prussia and Russia may have taken place long ago, but it has not been forgotten. This applies even more so to the Molotov–Ribbentrop

Pact, one of the most disgraceful episodes in the history of diplomacy. Officially just a non-aggression pact, the secret additional protocol called for a large-scale redistribution of Central and Eastern Europe between the two totalitarian powers. The new German–Soviet demarcation line ran from the Baltic to Bessarabia. At its heart was the mutually agreed dismantling of Poland and the disappearance of the Polish state once and for all from the map of Europe. The Pact gave Hitler free rein to attack Poland and France, while Stalin came closer to his goal of developing the Soviet Union into an empire that would surpass that of the powerful tsarist realm. The alliance with Hitler's Germany was explained by Stalin on 30 November 1939 in *Pravda*, the mouthpiece of the Communist Party of the Soviet Union, in his characteristic style – military terseness and a mocking undertone. The Great Leader of the Soviet Union stated:

1 that it cannot be denied that it was France and England that attacked Germany and consequently they are responsible for the present war;
2 that Germany made peace proposals to France and England, proposals supported by the Soviet Union on the grounds that a quick end to the war would ease the situation of all countries and peoples;
3 that the ruling circles of England and France rudely rejected Germany's peace proposals and also the attempts by the Soviet Union to achieve a quick end to the war. Those are the facts.

The message was clear: Great Britain and France were responsible for the Second World War, while Germany and the Soviet Union wanted peace. At this time the division of Eastern Europe between the two 'peaceful powers' had already taken place. On 19 September, when the Polish army was already defeated, Soviet troops entered eastern Poland. Just three days later, tank general Heinz Guderian and brigade commander Semyon Moiseevich Krivoshein attended a German–Soviet military parade in Brest-Litovsk. The commandant of the Red Army congratulated the Germans on behalf of the Soviet leadership on their military success. He promised to welcome the Germans in Moscow after their imminent victory over Great Britain. The Soviet leadership at the time could have no inkling that less than two years later German troops would be marching on Moscow. For Russia the Second World War began on 22 June 1941. For the countries of Central and Eastern Europe, it began with the pact between Hitler and Stalin in August 1939. For them, liberation from the Nazi terror also marked the beginning of a new reign of violence.

The kleptocracy no one speaks about

Almost no one wants to admit that the Russian power elite is enriching itself hand over foot. No one raises an eyebrow at the fact that Russian state prosecutors, members of parliament, Kremlin functionaries and finance officials own luxurious properties in London, Munich, St Moritz or the Côte d'Azur and move gigantic sums via phantom companies. The billions paid into slush funds during the building of the Olympic facilities in Sochi are of no interest to anyone. Nor does the Committee on Eastern European Relations, whom one would think to disapprove of the endemic corruption in Russia and the absence of legal security, lift a finger. Agreements are made and business is secured through arrangements and personal relations with the power apparatus. The majority of the German leadership elite ignore the fact that Russia is a kleptocracy in which participation in state power is a licence to make money.[1]

Putin's defenders on the left also appear unimpressed by the self-enrichment of the power elites and the crass social inequality. In terms of the gap between rich and poor, Russia is at the top of the list, well ahead of the United States.

The question of the Russian president is particularly awkward. Putin's links with mafia networks can be traced back to the early 1990s, when the political rise of the former KGB officer began in St Petersburg. When the Panama Papers in early 2016 revealed the worldwide network of tax evasion and money laundering, the name of the cellist Sergei Roldugin, a youthful friend of Putin, also cropped up. Transactions worth two billion dollars were carried out in his name through various offshore companies. On state television, Roldugin confirmed that he held shares in a number of Russian companies. The profits from these shares, he said, were used exclusively to promote young musicians. Putin had already set the tone earlier: 'As far as I know, he is a small shareholder in one of our companies and makes a little money from it, but not billions; that's nonsense. Almost everything he earns is spent on buying musical instruments abroad and bringing them to Russia.'[2] It remains to question how a modest musician had access to billions hidden in a labyrinth of phantom companies. There is no evidence that the innocent Roldugin was a stooge for his powerful friend. As with other shady transactions in Putin's entourage, everything bounces off him. He does not have to fear investigations in Russia, and western governments are wary of using information to this effect from their intelligence services.

How should we deal with the president of a nuclear power who is a political leader and the head of a mafia system at the same time? An embarrassing question and one that is therefore avoided. We deal with Putin as head of state and ignore the mafia boss of the same name. We have to deal with him – there's no doubt about that – but we should know whom we are doing business with. The dual nature of the regime is not unimportant when it comes to identifying the motives behind Russian politics. It's also about safeguarding 'jobs for the boys'. To praise Putin as an 'unblemished democrat' is a further distortion of all of the measures emanating from Moscow. Autocracy is democracy, attack is defence, the quest for liberty is fascism, the bombardment of cities is the fight against terror, lies are truth – welcome to Orwell's world.

Double standards in foreign policy

When Israel responded in July 2014 to a week-long hail of missiles from the Gaza Strip with a military counter-offensive in which, according to UN estimates, around 1,800 people were killed (most of them civilians), German public opinion was outraged. There were protests and demonstrations, and the German media carried daily reports. By comparison, the criticism of the Russian bombardment of Syria was mild, to say the least. The German peace movement seems indifferent to the fact that the Kremlin is supporting the despot Assad, that civilians are starving and that whole cities have been reduced to rubble. The Russian intervention in Ukraine has received a similar reception. The West could not simply ignore this obvious violation of European peace, but both the left and the right showed an understanding for Putin's actions, which fly in the face of international law. The majority find it difficult to take sides, much less to become involved in the conflict.

One constant feature of the German–Russian relationship is the deep-seated fear of drifting into a new major war with Moscow. The Kremlin understands perfectly well how to manipulate this. While Putin does not hesitate to use military might to gain strategic territories, this option is not open to the West as long as it threatens conflict with Moscow. This pattern was already seen in 2008 in Georgia and is now repeating itself in Ukraine and Syria. Russia presses ahead and creates a fait accompli through military force while the West shuns confrontation.

This is reminiscent of Karl Marx's description of the Russian–European asymmetry during the first Crimean War, when tsarist Russia proceeded to expand its empire at the expense of the Ottoman Empire:

> Counting on the cowardice and apprehensions of the western powers, [Russia] bullies Europe, and pushes his demands as far as possible, in order to appear magnanimous afterward, by contenting himself with what he immediately wanted. [. . .] The Russian bear is certainly capable of anything, so long as he knows the other animals he has to deal with to be capable of nothing.[3]

For Marx, tsarist Russia was the stronghold of European reactionism. Today, Russia under Putin has assumed this role again. In European public opinion, however, Marx's philippic against the western appeasement policy is unlikely to find much sympathy. Citizens and governments alike are agreed that the Russian bear should not be baited. In view of Russia's military arsenal and the danger of a nuclear war, there are indeed good grounds for a cautious approach to Russia. However, this does not mean accepting a redistribution of Europe into imperial spheres of influence out of fear of military threats. Europe cannot back down from the principles of Helsinki and Paris: renunciation of the use of force, territorial integrity, equal sovereignty and the non-alignment of all states. As much deterrence as necessary, as much cooperation as possible, without corrupting our values – that would be a suitable answer to Putin's revisionist policy.

Aims of Russia's German policy

Germany's infatuation with Russia is the reverse side of the anti-western stance of both national conservatives and virtuous leftists. The political strategists in Moscow know this weak spot very well. There is in any case a feeling that they often know the mood of Germany better than we know it ourselves.

Germany has always been a key country in Russia's foreign policy ambitions. The network of pro-Russian media, associations and institutions is correspondingly dense. The Russian embassy is an active lobbyist. There is no shortage of money, and if necessary Gazprom or Kremlin-friendly foundations function as sponsors. The cultivation of the political landscape ranges from the Linkspartei, which has transferred its loyalty seamlessly from the Soviet Union to Putin's Russia, to the Junge Union and the right-wing Alternative für Deutschland. The EU is deprecated. Since Putin's anti-western shift, which he announced at the Munich Security Conference in 2007, Russian politics has aimed at a new order in Europe, reducing the American influence, undermining the cohesion of the EU and extending the Russian sphere of influence. Within this sphere, the doctrine of 'limited sovereignty' applies once again. This is what is being demonstrated in Ukraine and is also meant to serve as a warning to others.

The wretchedness of the 'Russland-Versteher'

To obtain an impression of the particular mood in Germany – a mixture of guilt feelings, Russia kitsch, anti-Americanism and ignorance about the European nations that have been squeezed for centuries by Russia and Germany – one need only visit one of the many

German–Russian dialogue forums. They proclaim the virtues of German–Russian cooperation, rant about western sanctions and criticize the unholy influence of the Americans. One particularly popular credo of this community goes, 'If Russia and Germany understand one another, Europe is in good shape.' Anyone saying this can be sure of an enthusiastic reception. Almost no one will stand up and point out that this credo elicits completely different associations in Poland, the Baltic or Ukraine.

Putin's defenders in the West are frequently called 'Russland-Versteher' (literally Russia understanders). They do not deserve this epithet. Those who argue that Russia cannot be measured by 'western standards' are saying that the country and its population are incapable of democracy. They accept the narrative that Russia needs to be governed with a stern hand. History as destiny: because Russia was an authoritarian empire for centuries it must remain so in future. This turns the Russian citizen into a species unsuited for freedom. In other words, the so-called understanding of Russia's specificity is basically a *völkisch* cliché.

It goes without saying that in times of crisis, communication should be sought with Russia, both the power hierarchy and civil society. But dialogue is not the same as currying favour. The Russian authorities will certainly have a fine feeling for this difference. They react to weakness with disdain. They therefore deserve a clear message: Germany remains anchored in the West. It will not return to the see-saw policy of changing coalitions. We want good, equal relations with Russia but on the basis of the Helsinki principles, the renunciation of violence and equal sovereignty of all European states. And we will not remain silent when Russian human rights defenders, journalists, bloggers, environmentalists and feminists are gagged.

10

Modernity and Its Discontents

In 1818, a novel was published that still fires our imagination: *Frankenstein, or the Modern Prometheus* by the British author Mary Shelley. It is a product of the collective unconscious of the modern era. The story is quickly told: Viktor Frankenstein, a talented and ambitious young man, is obsessed by the idea of creating a man-made being. The combination of alchemy and modern science reveals to him the secret of bringing the dead back to life. After months of feverish work in his secret laboratory, he realizes his master stroke. A giant being, half-man, half-monster, is brought to life. Frankenstein's exhilaration turns to horror. He flees from his creature. The triumph of science turns into tragedy. The man-made creature turns against its creator. Rejected by everyone whom it attempts to approach, lonely and vengeful, it becomes a serial killer. Frankenstein also dies in the desperate hunt for the monster. Over the dead body of its creator, the unhappy creature has a realization and vows to kill himself.

In this story, Shelley linked two long traditions of modernity. First, she anticipated the scientific breakthrough in understanding the secret of life, advancing from research to the creation of living organisms. In its early stage, it concentrated on the systematic breeding of plants and animals; in its most recent guise, it involves genetic engineering and synthetic biology. At the same time, the social sciences were beginning to talk about the influence of society on individual behaviour. Personality and actions are neither God-given nor biologically determined but are shaped by social circumstances and experiences. Frankenstein's monster is both: the product of a bold inventive spirit that also breaks the last taboo, and an unhappy creature that mutates into a demon because its quest for love and fellowship is unrequited. Shelley's story may thus be interpreted in two ways: as a warning against human hubris that has the audacity to imitate God; and as a plea for compassion towards others affected by destiny. The sin in the first case is the creation of a man-made creature, in the second the rejection by its creator of a malformed creature.

The warning not to challenge the gods is one of the oldest myths in humanity. Shelley calls her human engineer Frankenstein a 'modern Prometheus', a contemporary version of the creator of man from Greek mythology who paid a terrible price for rebelling against the gods. In the Bible, the trouble starts when Adam and Eve defy God's command and taste the fruit of the Tree of Knowledge, forfeiting their innocence in this way. The Fall is followed by the banishment from Paradise, after which humans are born in pain and have to struggle to live their lives on earth.

Mankind continuously oversteps boundaries and is plagued by nightmares. We build skywards, redirect rivers and move mountains, fly around the globe at the speed of sound, rain thunder and lightning on enemies in war, communicate across oceans and continents and see pictures of events in the farthest corners of the globe, dissect life into its tiniest components and put it together again – all things that in the ancient sagas were in the realm of the gods.

Now we are about to overstep another boundary: from the reshaping of nature to changing ourselves. In addition to the ancient self-help disciplines – physical training, spiritual exercises, lifelong learning – there are now new processes that encroach far into our physical and intellectual constitution. Psychopharmaceuticals, performance-enhancing substances, artificial insemination, cell therapy and cosmetic surgery have long pointed in this direction. We are on the threshold of human engineering, self-optimization through medical technology. It is driven by the age-old desire for eternal youth and beauty, for increasing our intellectual and physical capacities, for new experiences. We don't know where this will all lead. We won't win the race with death. But the journey into the unknown that began with the banishment from Paradise is by no means over.

Scientific knowledge grows exponentially. But this increased knowledge also gives us greater possibilities for good and for evil. How can we limit the risks without sacrificing the benefits of scientific and technical progress? Research requires normative limits. It is an ethical precept that human life should not be used as a means to an end. But there is a wide field on this side of the boundary in which the considerations that need to be taken into account are by no means black and white. The most important insurance against inhuman developments is an informed critical public. The greater the impact of new research and technology, the more it must be placed in the spotlight.

Goethe's *Faust*: ruin comes on running feet

Viktor Frankenstein is a soulmate of Dr Faustus. They both come from a time of revolution in the early nineteenth century, a period of transition from the old to the new world. The long era of feudalism, estates-based society and rural life was coming to an end. Rapid developments in science, technology and industry, the growth of cities, the rise of the bourgeoisie, demands for freedom and equality heralded a new era. The generations of that time no doubt had similar feelings as those today. They were caught in a whirlwind that blew away everything they were familiar with and spun them into an uncertain future. Goethe's *Faust II*, on which he worked until his death in 1831, captures this upheaval in a great human drama.

With our modern eyes, sharpened by crises and disasters, the work reads like a melancholy farewell to the old world and a clairvoyant warning against the demons of the new one. Faust is reminiscent of the modern go-getter, restless and ruthless in the search for power, wealth and immortality. His burning desire to know 'whatever binds the innermost core of the world together' is not an end in itself. Knowledge is the instrument for subjugating nature, a means for new ventures. He is an adviser to the emperor, financial magician, entrepreneur, warlord. Every success merely arouses new ambitions; every step leads to further steps. Never does he arrive at a moment of happiness when he could say 'Stay awhile, you are so lovely!'. Faust the doer is a driven man.

In places, *Faust II* is like an original poetic version of an ecological manifesto. It contains practically all of the motifs of a modern critique of capitalism. The constant restlessness and incessantly accelerating tempo of life is diabolical,[1] the miraculous increase in finance by printing paper money is alchemical hocus-pocus. War, trade and piracy are notorious companions of capitalism, 'as three in one, no separation'. The expansion of industrial production demands the plundering of natural resources. It leads to the destruction of entire regions, the loss of original beauty and harmony. The more wealth grows in the hands of the propertied class, the greater the worry about losing it. The 'three mighty warriors' that help Faust in his ventures are translated into English as Bullyboy, Grab-Quick and Hold-Tight – representing violence, greed and avarice. We acquire with violence, want more and more and hold on tightly to what we have gained.

What others describe as historical progress – the unleashing of science and technology, the increasing control of nature, the delimitation of space and time – Goethe sees as a pact with the devil. The downfall of the modern age is already pre-programmed by its achievements. As the high point of his life work, Faust embarks on a large-scale land acquisition plan. He seeks to turn the sea into a settlement area for millions. 'Free earth: where a free race, in freedom, stand' is his last great vision. It is a project completely in the spirit of the modern times: engineering skill and hard work to control the forces of nature. Mephisto knows better. He sees behind the newly erected dykes and canals a great wave building up: 'And yet with all your walls and dams / You're merely dancing to our tune: / Since you prepare for our Neptune, / The Water-demon, one vast feast. / You'll be lost in every way – / The elements are ours, today, / And ruin comes on running feet.'

And this is precisely what happens. When the blind Faust hears the clattering of shovels at the end, he imagines the army of labourers constructing his great work. He thinks himself close to immortality: 'Through aeons, then, never to fade away.' In fact, what he can hear is lemurs digging his grave. At least he doesn't have to redeem his wager with the Devil. Angels carry his soul to higher spheres: 'Whoever strives, in his endeavour, we can rescue.' In the end, it is not his works that save Faust from damnation but love: 'Woman, eternal, beckons us on.'

No, the modern world is not Goethe's world. He regarded its progress with aristocratic melancholy. Industrial capitalism was still in its infancy, but Goethe already saw the destructive force connected with finance and the industrial revolution. He left no doubt that the original accumulation of capital was a distressing process connected with violence and war. His scepticism went even further, however. For him, modern man was a sorcerer's apprentice who changes the world without being able to foresee the effects of his experiments. The spirits that he summons get out of control. He imagines that he can control nature but merely digs his own grave. In his quest for power and wealth, he loses his sense of the beauty of nature and the ability to enjoy life. This was a sharp view of the dark side of the modern age. What Goethe didn't see (or did not wish to describe) was its bright side, the side of light.

The disappearing world that Goethe mourned was anything but a bucolic idyll. For the vast majority of people, life was short and hard. Nutritious diet, education, art, travel, comfort and leisure were a privilege of a tiny upper class. The mass of the rural population was exposed to the caprices of nature and the despotism of the authorities. The tranquil life was permanently threatened by sickness, bad harvests and compulsory levies. Millions of German rural inhabitants emigrated to America to escape the social hardship and to be able to breathe freely. It was not until the development of science, technology and industry that a hitherto unseen increase in life expectancy, education and social security was achieved. The danger of man-made disasters still exists and has increased since Goethe's time with the growing impact of modern technology on new dimensions of life. But a 'back to nature' is possible at best for a small group of dropouts, but not for most of the seven billion people, soon to become nine or ten billion, inhabiting the earth. The only way forward is towards a reflective modern world that doesn't stumble blindly onwards but offers room for criticism and correction.

Malthus and the limits of growth

One of the credos of the green community is that unlimited economic growth is not possible on a limited planet. This is no more than common sense. We don't have a second planet in reserve, so we have to make do with what we find on this one. There is a limited quantum of land, natural resources and water. The ability of ecosystems to absorb pollutants of all kinds is also limited. If we emit too much carbon dioxide into the atmosphere, the earth heats up. If our consumption of natural resources is greater than their rate of reproduction, if, for example, we pump more groundwater out of underground reservoirs than can be renewed, the supply dwindles. We will then be using up the planet's substance. At some point, the natural resources will be exhausted, leading to the collapse of entire ecosystems and the foundation for human life. Such over-exploitation of natural resources can already be seen in many parts of the world. The overgrazing of land leads to desertification; deforestation or the burning of forests produce bleak moonscapes; agricultural monocultures leach the soil; the overfishing of the seas threatens the survival of entire species. If the maximum ecological limits are consistently exceeded, we may end up in an unremediable situation that can no longer be saved.

Can the threatened ecological collapse be avoided only by a radical, self-imposed limitation of humanity's birth rate, production and consumption? One of the primogenitors of this way of thinking was the British theologian and economist Thomas Robert Malthus, a contemporary of Shelley and Goethe. His fatalistic population theory, that only a drastic limitation of the birth rate could rescue mankind from catastrophic crises, still haunts us today. The starting point for his theory was the seemingly immutable law that agricultural production shows linear growth while populations grow exponentially. In other words, agricultural production cannot keep pace with human production. Supply and demand diverge increasingly. The inevitable consequence is famine and wars fought over scarce agricultural land. They will periodically decimate the population and restore the 'natural balance'. The only way of breaking this vicious circle is to reduce the birth rate. As the supply of food can be increased by only a very limited amount, demand must be adapted to the scarce resources. This remains the prevailing theory, far beyond the environmental movement.

Malthus's ecological determinism was refuted by real developments. On the basis of the agricultural productivity of the time, he concluded that the earth could just about feed a billion people. This was about the population of the earth at the time. It has now increased sevenfold, the per capita calorie consumption has risen around 50 per cent and life expectancy is twice as high as it was at the start of industrialization. At the same time, the proportion of the population involved in agriculture has dropped drastically. At the beginning of the twentieth century, agricultural workers still accounted for around 60 per cent of the population of Germany compared with just 3 per cent today. The key to this 'loaves and fishes' miracle is in the enormous boost in agricultural productivity. In contrast to Malthus's premises, agricultural production rose much more rapidly than population. Like all models based on fixed natural limits for production and consumption, he underestimated a decisive factor: human inventiveness. Knowledge-based innovation is the greatest productive force of all. In this case, it was the invention of mineral fertilizer by Justus von Liebig, combined with refined methods of animal husbandry and plant cultivation, the use of agricultural machines and better transport links between the country and the city. All this increased production by unsuspected amounts and reduced losses in cultivation and on the way to the consumer. Even today, agricultural potential is far from being exhausted. Recultivation of the soil, improved planting methods, land reforms, progress in plant breeding, better education, development of the rural infrastructure in developing countries, more plant-based food instead of more feed for meat production – if the groundwork is done properly, even with a population of nine or ten billion by the mid-century, no one needs to go hungry. We are far from reaching the limits of our possibilities on this planet. At the same time, this should not be seen as a licence for 'more of the same'. The risk of serious ecological crises is real enough. However, as Friedrich Hölderlin said so appositely: 'But where there is danger, a rescuing element grows as well.' We can transmit satellite pictures from Mars and clone living beings, build supercomputers and transplant human organs. Why shouldn't we manage to decarbonize the energy system? This is not a question of technical feasibility but of political will.

Marx and the modern world

Just a few years after the posthumous publication of *Faust II*, Karl Marx entered onto the historical stage. He offered a completely different tone. Where Goethe had described the disappearance of the old world as a disaster, Marx welcomed the rise of capitalism with unconcealed enthusiasm. There is hardly a more penetrating description of the revolutionary, restless and unbounded character of the new epoch than in the *Communist Manifesto* in 1848:

> Constant revolutionising of production, uninterrupted disturbance of all social conditions, everlasting uncertainty and agitation distinguish the bourgeois epoch from all earlier ones. [. . .] All that is solid melts into air, all that is holy is profaned. [. . .] In place of the old local and national seclusion and self-sufficiency, we have intercourse in every direction, universal interdependence of nations. And as in material, so also in intellectual production. The intellectual creations of individual nations become common property. National one-sidedness and narrow-mindedness become more and more impossible, and from the numerous national and local literatures, there arises a world literature. [. . .] Subjection of Nature's forces to man, machinery, application of chemistry to industry and agriculture, steam navigation, railways, electric telegraphs, clearing of whole continents for cultivation, canalisation of rivers, whole populations conjured out of the ground – what earlier century had even a presentiment that such productive forces slumbered in the lap of social labour?

Who today would celebrate the achievements of industrialism with such unbroken enthusiasm? Where Goethe mourned the idyll of rural life from his aristocratic standpoint, Marx welcomed the rapid growth of cities, which 'has thus rescued a considerable part of the population from the *idiocy of rural life*'. He celebrated the suppression of local businesses by major industry and the globalization of production and consumption as progress. World trade placed new demands that were no longer bound to domestic resources. Provincial narrow-mindedness was being overcome. The mutual dependence of nations fostered the development of a global society built on trans-border circulation of ideas, goods and people.

Like Goethe, Marx also predicted the transition to this new society formation, not because it overstepped the bounds set by nature and the gods but because the permanent revolution of the means of production stretched capitalist ownership to its limits. Capitalism would fail as soon as the productive forces it required went beyond the narrow ownership situation. The industrial technological revolution was great as it paved the way towards a classless society. The great leap from the realm of necessity to the realm of freedom was conditional on the all-round development of productive forces in society.

The prevailing criticism of capitalism is the inverse of Marx's panegyric on the revolutionary character of the 'bourgeois epoch'. It fears precisely what Marx praised as the historical merit of capitalism: the limitless *development of productive forces*. In contrast to the pure Marxist doctrine, the rate of scientific and technological innovation in modern capitalism is accelerating. It continues to infiltrate new areas and penetrate deeper and deeper into our lives. On the one hand, the creation of artificial life through breakthroughs in genetic research and synthetic biology are within grasp, while on the other globalization of our fossil economic system threatens the stability of the earth's climate. Both produce a reactive force. Today's criticism of capitalism is closer to Goethe than to Marx. It aims not at setting free productive forces but at slowing down and limiting them.

Where Marx attacked the protectionism of Prussian country squires, the left today is at war with free trade. Where Marx praised permanent technical revolution as the progressive side of capitalism, the Greens see new technologies primarily from the point of view of damage control. Scientific and technological progress has lost its innocence. Mephisto's verdict 'ruin comes on running feet' has been given a new dimension with climate change.

But we should not throw out the baby with the bathwater. The ecological and social challenges of the twenty-first century cannot be met with an unreal withdrawal from modernity. They demand not a slowing down but a *speeding up* of technological change, not the subordination to nature but the targeted development of its potential.

Sloterdijk: modernism as a disaster

Peter Sloterdijk is the uncrowned German king of philosophy. He would be a sure-fire medallist in heavyweight popular philosophy. He possesses an almost unparalleled talent as a zeitgeist medium. His theory of modernity reflects the basic pessimistic view of civilization, which sees its achievements as also being at the root of its decline. Sloterdijk's topical statements on the refugee question, his warning against the surrendering of control by the state and the relinquishing of national sovereignty are derived from his interpretation of modernity as a process of social entropy, the steady loss of social and political order. If modern liberalism and its promises continue to produce a surfeit of expectations, hopes and demands that cannot be met, then it will be important above all to strengthen institutions promising stability and a capacity to act – hence the rehabilitation of borders as a response to the trend towards their steady dissolution.

Sloterdijk's 2014 work *Die schrecklichen Kinder der Neuzeit* (The Terrible Children of the Modern Age) sets a new culturally conservative tone. As usual, he spreads himself wide. He offers an alternative to the Marxist interpretation of the 'social question' as a historical driving force behind the development of civilization, namely the 'genealogical question', the interpretation of the history of humanity from the dynamic of *heredity* and *generation breaks*. Heredity stands for the continuation of traditional values and lifestyles, generation break for refusal, revolt and a new start. While for the longest time human development was based on the premise of maintaining the status quo, the modern age is notable for the abandonment of tradition, the interruption of generational continuity and the cult of the new: new is better, old is backward. The party of progress triumphs over the party of conservation, activism triumphs over conservatism: 'What exists and persists will be wrong; what goes forward and marches for freedom will have right on its side.'

It is the 'terrible children' of their epoch that are the agents of subversion, 'glorious bastards' who compensate for the stigma of their origins with high-flying ideas and missionary zeal. Sloterdijk follows the traces of some of these personalities in the history of the world, starting with Jesus of Nazareth and Napoleon Bonaparte: illegitimate children, neglected heirs, outsiders from birth, who make a virtue out of necessity. Until the beginning of the modern era in Europe, such individuals were isolated cases, but appeared thereafter en masse until (according to Sloterdijk) they became the prevailing type in modern western societies: people without shadows, who do not see themselves in the continuation of family and cultural traditions but as 'unattached' in the deeper sense of the word.

Individualization means liberation from traditional conventions and hence from the binding obligation to previous or future generations. If the nineteenth century was the age of forward-looking projects, technical and political utopias, the postmodern era is stuck in the here and now. We have broken free of our origins; the future is clouded with the threat of conflicts, national debt and climate change. The confidence in progress of the modern period is giving way to a *fin-de-siècle* mood. Even in the power elites, there is the feeling that western capitalist civilization is past its zenith; in future, the only direction will be down. Sloterdijk is a prophet of this sentiment.

Doomsday mood

In the make-up of society today, Sloterdijk sees confirmation of Nietzsche's 'last man': 'end consumers' of the goods and relations that make up the wealth of society. Their denial of the future culminates in a refusal to reproduce, the lowering of the birth rate below the level required for society to reproduce itself (*Schrumpfvergreisung* – shrinking ageing). Two eighteenth-century women serve as oracles: Louis XV's mistress Madame de Pompadour with her infamous 'Après nous, le déluge', who anticipated the twilight of the aristocratic era, and Laetitia Ramolino, Napoleon's mother, who remarked on the ambitious rise of her son by saying 'Pourvu que cela dure' (let's hope that it lasts).

These two women reflect Sloterdijk's thinking exactly. In conclusion, he offers a glimmer of hope for the future. 'It is not yet clear whether the final figure will decide to continue for the time being or whether it will follow the inclination towards a pyrotechnic ending in the here and now.' But the basic tenor of his views of history is extremely sceptical. The power thinker Sloterdijk reveals himself to be a cultural conservative. Progress is in quotation marks – as the road to doom. The epoch after 1789, coinciding with the industrial revolution, is seen in retrospect as a series of disasters. And so it was. But behind the terrors of totalitarian regimes, the bloodshed and atrocities of wars that have laid waste to Europe since then, another picture disappears: one of breathtaking social progress and political emancipation associated with modernity.

Global economic performance has exploded; the standard of living, education level and life expectancy of billions of people have risen to unimagined heights. New groups continue to claim equality. Modern societies offer space for a hitherto unseen diversity of lifestyles.

The number of people in the world in extreme poverty has declined in both absolute and relative terms, as has child mortality and illiteracy. Does that mean that everything is fine? Of course not. An alarmingly large number of people still live in wretched circumstances. Their lives are dictated by hunger, violence and the struggle for survival. Looking back on the last two centuries, however, it makes an important difference whether one regards the history of modernity as a disaster or a time of progress accompanied by crises and setbacks.

It would be wrong to suggest that Sloterdijk's worldview is a linear one. He is a master of ambivalence and paradoxes. Again and again, he alludes to modernity as an emancipatory project, a process of progressive political and social participation of new and larger social groups. In contrast to the conventional opinion that we are confronted all the time by new forms of exclusion, he suggests that the French Revolution marked the start of a dynamic of *increasing inclusion*: the rise of the third estate (the bourgeoisie) was followed by the abolition of slavery, the emancipation of the workers, of women and of the colonies. The proclamation of human rights and the normative ideas of liberty, equality and fraternity developed an unstoppable momentum of their own that could no longer be controlled by their authors or anyone else. Sloterdijk therefore calls the modern era the 'era of demands with no end in sight': new groups continue to demand equality and participation.

Social entropy

He describes the civilization process as one of increasing disorder. The liberated individual ambitions and social energies have outgrown the diminishing cohesive effect of social institutions. Sloterdijk speaks of an unfulfillable inflation of aspirations unleashed with the expectation of equality: demands for social advancement, access to lucrative positions, public attention, luxury. In estate-based societies, social position was established through birth. Origins determined opportunities in life, and individual life generally remained on a predetermined path. This genealogical determinism became more relaxed in the transition from the Middle Ages to the modern era. In place of a fate predetermined by birth came the idea that anyone could become (or have) anything. The era of exclusive pre-emptive right to offices, power, prominence and luxury was followed by a 'stampede for positions in a life without preconditions'. Modern societies are fast-moving dream machines. The tension between limitless expectations and limited opportunities for realizing them explains the querulous, disappointed undertone in public sentiments, in spite of hitherto unimagined prosperity.

Sloterdijk's key concepts for the modern era are *Freisetzung* (release) and *Entgrenzung* (delimitation). He tends to see them as a liability. The dissolution of the traditional order, passed on from generation to generation, will lead to an increase in social entropy. Borrowing from the second law of thermodynamics, he constructs a 'law of civilization dynamics', which he sees as the basic model for historical change: 'In the global process after the hiatus [Sloterdijk-speak for the widening gap between the generations], more energies will be released than can be bound under forms of transferable civilization' – more ambition, wishes, demands for participation, life options, potential for revolt, assertable rights, erotic desires, but also more credit, emissions and other residual waste of all kinds that will encumber future generations. In short, the accumulation of problems will exceed the ability in future to resolve them and the released social energies will exceed the 'ability of cultivating cohesive forces'.

Referring to Marx, but without the optimism of his philosophy of history that crises and contradictions will give rise to a higher order, Sloterdijk describes modern capitalism as a process of 'destabilization of all relations, both material and symbolic'; a 'permanent tumult of revolutions and deracination [will replace] stable cycles.' One of the fatal consequences of this leap into the open is the release of excessive violence, which Sloterdijk sees in the modern era. In this light, the French Revolution was not a liberating act but the mother of all disasters to befall Europe. For Sloterdijk, it was the origin of a new 'alliance between the human and the infernal'. 'People are never more obsessed in their actions than when they are filled with an awareness of their freedom.' The most terrible acts of violence are committed under the noblest pretexts.

Redemption and barbarity

According to Sloterdijk, the modern period saw the start of an 'era of waste of human life'. He quotes Napoleon: 'A man like me troubles himself little about a million men.' He was the pioneer of mass mobilization, mass warfare, mass killing, which was continued in the twentieth century in even greater style. But were the puritanical, religion-drenched societies of the old (and modern!) era less malevolent, less cruel, less bloodthirsty? Were there no show trials, torture, pogroms, devastation of entire regions, massacres of women and children? Were they more respectful of life? The opposite is probably true: human rights and international law are modern ideas. There is indeed something to be said for the view that the modern era has been one of gradual containment of arbitrary violence, both in everyday life and in the interaction between states. The destructive force set free by the First and Second World Wars was rather the result of modern forces of production (war technology) than a demonstration of a new quality of degenerate brutality. The carnage of war and the annihilation of an entire people now took place on an industrial scale.

What followed these orgies of destruction were political attempts to prevent their repetition through the establishment of legal norms and supranational institutions: the League of Nations and international law, the Geneva Convention, arms control, the responsibility to protect as a provision of international law. The fact that these provisions were anything but a storm-proof guarantee for 'never again' is another matter. The process of civilization remains threatened by the possibility of a relapse into barbarity.

According to Sloterdijk, the progress of modernity is a 'chronic surge forward camouflaged as deeds, projects and planned actions'. This applies at least to the capitalist economy, the 'global interior of capital' that is spreading all over the world. Its guiding principle is more and faster, every interruption to this dynamic leading to crisis. The spaceship is out of control and is heading on an uncharted course without a recognizable destination or landing place. It is the basic feeling of drift articulated by Sloterdijk: we are passengers in a mega-machine that we have long lost control of.

A central challenge facing modern society is how to strengthen its control of the rapid freewheeling of the economic and technical dynamic. One way is to expand the action horizon of politics and businesses from the short to the long term. Sloterdijk calls 'sustainability' an 'auto-hypnotic formula'. It reflects the conservative longing for permanence. 'To rediscover the forgotten art of permanence' is in many ways a good thing, from the approach to nature to a mature approach to relationships. We need elements of continuity to be able to master the change: long-term personal relationships, robust institutions and a normative basic consensus that survives changing governments.

11

Ecology and Freedom

> There is no return to a harmonious state of nature. If we turn back, then we must go the whole way – we must return to the beast.
>
> Karl Popper

In principle, we have been aware for years of the critical nature of the ecology question. Climate change, the ongoing loss of fertile land and the threatened collapse of maritime ecosystems are the ingredients for an epochal crisis that will shake the twenty-first century to its foundations. They force millions of people to leave their homes, spark conflicts for scarce resources and destabilize entire regions. Ecological crises mean that our lives are dictated by outside forces rather than ourselves. And yet ecology and freedom are not twins, just as nature conservation is not the sole preserve of 'progressive forces'. In historical terms, the opposite tends to have been the case. The reverence for rural agriculture was a conservative romantic elegy; complaints about the alienation of city dwellers from nature, the rejection of commercialization and mass consumption and the association of civilization with degeneration came from the right. Vegetarianism and biological cultivation, homeopathy and anthroposophy have never been progressive. They are ingredients of both *völkisch* fascist thinking and its opposite.

There needs to be a conscious decision for a liberal ecology policy. In the early history of the Greens, this was by no means agreed. Among the German Greens, left-wing bourgeois liberal and conservative spirits encountered blood-and-soil environmentalists from a completely different planet. They all fulminated together against the pollution by the chemical industry and food companies, against nuclear death and NATO, called for a 'Europe of regions' and saw themselves as fundamental opponents of the 'prevailing system'. This allowed plenty of scope for movements with opposing sociopolitical ideas. It was only after a whole series of raging internal conflicts and separation processes that the Greens became what they are today, a left-of-centre reform party. Decisive elements on the way were the discussion on abortion and women's right to self-determination, a coming to terms with parliamentarianism, the bitter dispute on participation in government and, not least, the conflict regarding NATO and Germany's commitment to the West.

The historical merit of the Greens is in their linking of ecology and liberal democracy. When it comes to civil rights, sexual choice and diversity of lifestyles, today's Greens are in their element. In other areas, liberal green politics has yet to prove itself. When ecology and economy are at stake, the anti-authoritarian legacy of the Greens collides with the equally deep-seated temptation to decide what makes people happy. Here the temptation of state control looms: the state should fix and regulate everything on both a large and a small scale.

As the unsuccessful attempt to impose a compulsory meat-free day in schools and other public institutions ('veggie day') showed, the public reacts sensitively to state intervention in private life. In spite of all attempts to become less rigid, the Greens still have the image of a ruling party that wishes to guide its subjects towards the path of righteousness by means of orders and prohibitions.

The conflict between individual liberty and ecological imperatives is by no means just a green obsession. The founding document of the ecology movement, published for the Club of Rome under the title *The Limits to Growth* in 1972, has an underlying authoritarian tone. As a response to the threatened self-destruction of industrial modernity, the authors suggest comprehensive control of production, consumption and reproduction. In place of the market and competition comes the ecological authoritarian state.

Dennis Meadows, flag bearer of *The Limits to Growth*, still does not believe that parliamentary democracy is capable of moving away from the egoism of short-term interests in favour of the long-term interests of the species. As he sees no alternative to a massive restriction on consumption, this scepticism is not unfounded. It is difficult to imagine parliamentary majorities being persuaded to support a reduction in earnings and consumption.

His long-standing comrade-in-arms Jørgen Randers goes a step further and sympathizes openly with Chinese authoritarianism. He idealizes the politburo of the Communist Party as a 'benevolent dictator' that can implement necessary ecological measures with a strong hand. Only a strong central power that can override property rights and individual interests can with one stroke of the pen close a hundred obsolete paper mills and replace them with a huge factory operating according to the latest environmental standards: 'Such decisions are of long-term benefit to the environment and would be difficult to implement in a democratic society.'

His admiration is shared by many western managers bothered by the slowness and need for compromise in democracy. Stability and 'efficient governance' are more important than human rights and political freedom.

China's successful economic development and the advancement of hundreds of millions of people into the modern middle class appear to bear out this view. But even if we allow that a regime should be judged only on its success in modernizing the country, Chinese authoritarianism is not a good model. The combination of uncontrolled capitalism and one-party rule has catapulted China in a very short time from being a backward agricultural country to a modern industrial one. The price for this enforced industrialization has been and still is high. Media censorship, subordination of the judiciary to party organs, omnipresent corruption and the suppression of public protest make possible the most brutal pollution of air, soil and water, whose cost is not taken into account. At the same time, the unbalanced growth model based on the one-sided expansion of heavy industry has led to enormous distortions in the economic structure. After the hypergrowth began to abate, basic industries and the fossil fuel energy sector operated to just 60 or 70 per cent capacity. Chinese banks are sitting on a pile of bad loans. Regional disparities and social contrasts within China are enormous. It remains to be seen whether the country will succeed in its transition to an innovation-based and service-oriented economy within the current political structures.

The Party's power monopoly, the rigid control of civil society, the absence of the rule of law and the dominance of state-owned companies are obstacles to the next stage in economic modernization. China hopes to become the leading digital power. At the same time, state control of the internet and the tight censorship of social media are being stepped up. This raises the question of how much freedom a knowledge-based economy requires. The same applies to the transition to a resource-sparing, environmentally friendly economy. As long as environmental and consumer associations, media and the judiciary are kept on a leash by the Party, it will be difficult to observe the environmental protection regulations, which look impressive on paper, throughout the country. The ecological transformation is not merely a top-down project. It calls for transparent information, critical media, a vital civil society and an independent judiciary.

Authoritarianism in green

The ecological flirt with dictatorship is no coincidence. It sets the survival of human civilization against the freedom of the individual. With a little dialectics, the impending ecological disaster can be used as an excuse for authoritarianism in the name of freedom. Those who defend the natural basis for existence also defend the freedom of coming generations not to have to spend their lives under the diktat of rising temperatures and dwindling resources. What could be more logical than to restrict the freedom of the present generation by appealing to a humane future?

The idea of preventing future disasters can easily become a licence for paternalism and restriction. In the face of the threatened collapse of entire ecosystems, a wasteful lifestyle, free consumption and freedom to travel and do business seem like frivolous luxuries. What is so wrong about banning fuel-guzzling SUVs or strictly limiting plane journeys? When climate experts calculate that we cannot allow more than two tons of carbon dioxide per person per year to be released into the atmosphere in order to maintain global warming at under 2°C, why do we not give everyone emission quotas to govern their lives?

What appears justified from an ecological Jacobinist point of view would be a large step towards a police state of Orwellian proportions, in which every purchase, every trip, every steak would count towards a personal carbon dioxide account. It is understandable that those who see the ecological crisis as the result of excessive human demands should conclude the need for an authoritarian approach. This would mean in the first instance a restriction on consumption and the fair distribution of the smaller amount: a combination of eco-puritanism and authoritarian planned economy, which would give everyone their fair share and would watch carefully that cities, businesses and individuals did not overdraw their ecological accounts.

It is but a small step from saving the world through voluntary restraint to re-educating the modern consumption-obsessed individual. If the causes of the ecological crisis are seen in individual self-indulgence, it is only logical to seek the solution in mental reprogramming. Asked how we can face up to the ecological disaster, Dennis Meadows answers: 'This would change the nature of man. [. . .] My concern is that for genetic reasons we are just not able to deal with such things as long-term climate change.' The project for improving mankind has a long tradition. Its ascetic variant calls for self-cleansing through denial of everything that is unessential, all frippery and luxury. If people are not willing to abandon the sinful path of 'more and more', the authoritarian version of mandatory denial beckons.

In September 2016, a new report for the Club of Rome was published with the title *Reinventing Prosperity: Managing Economic Growth to Reduce Unemployment, Inequality and Climate Change*. Once again, Jørgen Randers was one of the authors. At the book presentation in Berlin, he came out with the utterance, 'My daughter is the most dangerous animal in the world' because she consumes thirty times as much resources as children in poor countries. Randers's choice of words was not accidental. It is in line with a worldview in which the ecological crisis is the reward for the self-indulgence of people in the prosperous world. For modern Malthusians, there are too many people in the world who produce and consume too much. The answer to climate change and the resource crisis thus lies in less: limiting the birth rate, shorter working hours and lower income, reduced international trade and restraining economic growth.

The latent misanthropic element in ecological thinking aimed primarily at *restriction* rather than *innovation* can be seen in the demand that women in industrial countries should be paid a premium of US$80,000 if they have no children, or at best one child. One might ask what problem this demand is meant to be solving. In almost all old industrial societies, the birth rate is in any case below the average of 2.1 per family required to maintain the population level. In developing countries as well, the birth rate goes down with increased education and a better standard of living. Those who want to protect women from having more children than they want should support women's rights, school education for girls and free contraception. In any case, hunger is not a result of population growth. The earth could feed ten billion people if the unequal distribution of land in many developing countries were corrected, if farmers were better educated, if more were invested in the rural infrastructure and if more human food than animal feed were produced.

Like the original 1972 report, the new version remains wedded to *quantitative logic* instead of seeing the roots (and solution) to the ecological crisis in the *way* in which we generate energy, build cities, organize industrial processes and manage natural resources.

With their penitential sermon, Randers and the others miss the most important point: the decoupling of value production and resource consumption through a green industrial revolution. What we need to limit is the squandering of resources, emissions and waste; what we have to encourage on a large scale is research, inventiveness and innovation. As Marcel Fratzscher, head of the German Institute for Economic Research, puts it: 'The solution is not having fewer children but making use of mankind's most important resource, its creativity and innovative strength.'

Because we are in a race with climate change, we need a higher rate of innovation for the decarbonization of transport, industry and energy production. This also applies to developing countries. The industrialization of Asia, Latin America and Africa on the basis of fossil fuels is causing an ecological collapse. It is not sufficient to grow with a little less damage to the environment. The alternative is to mutate as quickly as possible to a green economy based on renewable energies, environmentally friendly transport systems and waste-free material cycles.

Those, like the author of these words, who believe that climate change will be a central challenge in the next few decades back their statements with findings from climate research. There is a good reason for this. Knowledge of the climate is based on interdisciplinary research and complex modelling. Interested non-specialists have practically no other option but to rely on the majority consensus of the scientific community. If we are in any case reliant on science, there is a temptation to make it the highest instance that should tell us what to do. Politicians then become mere executors of what science tells them.

'The climate is non-negotiable' is a favourite credo within the green community. Everything is subordinate to combating climate change. The actions of parliaments, governments and citizens are guided by the results of climate research. Economic imperatives are replaced by ecological ones. This logic would lead to the abolition of politics as the forum for public argument between diverging views and interests and the consideration of competing aims. Both climate scientists and environmentalists should beware of such ecological absolutism. Scientists can highlight risks, sound alarms and offer recommendations. But climate research should not have priority over other scientific disciplines, nor should it pre-empt politics.

In any case, the prevailing doctrines are not always the gold standards for environmentalists. They refer to the scientific majority consensus only where it furthers their own agenda. The fundamentalist opposition to genetic engineering of plants, for example, ignores the scientific mainstream opinion. This is not to say who is right or wrong. The history of science is full of examples of how new findings had first to overcome the established majority view. To that extent, reference to the educated majority is no guarantee of the validity of any particular point of view. Reference to 'science' is not the ultimate argument in the discussion of political priorities. This also applies to climate policy.

In environmental politics as well, a noble end does not justify every means. Freedom is more than just an understanding of necessity. Democracy is a value in itself that should not be invalidated in favour of a green TINA ('there is no alternative') principle. It is not yet clear how a liberal environmental policy that does not succumb to the temptation of comprehensive control would look. No environmental policy can exist without regulations, limit values and prohibitions. But they are not the key to the solution of the environmental question. Our most important resource is human creativity – and in that regard as well democracies are a better alternative than authoritarian regimes.

Primacy of nature?

The difference between an authoritarian and a liberal interpretation of environmental policy starts with our understanding of the famous 'limitations for growth'. Is it an immovable barrier to what is possible for the inhabitants of our planet? In that case, environmental policy must make human society subordinate to the laws of nature. Given that logic, human civilization would be just a subdivision of the ecosystem. The physics and chemistry of the terrestrial system would be the true gods, and those who did not subordinate themselves to them would be punished by floods, drought, thunder and lightning.

Empirical experience and scientific research have demonstrated that there are invisible limits to the ecosystem that we should do our utmost not to exceed. At a certain more or less foreseeable point, the climate will get out of control, entire populations will collapse, fertile land will become desert and biological river and marine life will become impoverished. The important point is that no fixed limits for the development of human civilization can be inferred from these red lines in the ecosystem. Just as Malthus was wrong in the early nineteenth century in his prediction that the earth could feed a maximum of a billion people, the degree of prosperity possible for the nine or ten billion people that will populate the planet by 2050 cannot be deduced from ecosystem research.

The extent of the conflict between economic growth and ecology is first and foremost a question of the *method of production*. There are three fundamental changes through which the limits for growth can be extended: first, the transfer from fossil to renewable energies; second, a continuous improvement in resource efficiency (making more from less); and third, the transition to linked reusable material chains, in which every residue is biologically or industrially recycled. Together they form the guidelines for a green industrial revolution aiming to decouple value creation from the use of national resources. A liberal environmental policy is a question not of changing humanity but of introducing a new method of production – away from predatory exploitation of nature towards intelligent production in harmony with nature.

In classical liberalism, political freedom (democracy) and economic freedom (a liberal view of the economy) are mutually dependent. Within this tradition, the concentration of economic power in the hands of the state is a danger, and private property is the guarantee of civil liberty. Markets are a form of economic self-organization that links countless producers and consumers. They combine the knowledge, capabilities and preferences of many individuals. To that extent, they are superior in principle to any form of state economic control. For most Greens, this is an alien notion. They regard markets with mistrust, associate entrepreneurship with greed, and competition with ruthlessness. There are indeed many demonstrations of this. But this view fails to recognize the productive force generated by the market economy and entrepreneurial spirit.

Green ordoliberalism

'Green ordoliberalism' could be the way to balance state and market. Markets require a lot of conditions to function. They depend on factors that they cannot provide themselves: legal security, a public education system, efficient transport routes and data networks, protection against cartels and monopolies, and prices that accurately reflect the cost of the product. To a certain extent at least, the welfare state is also a condition for the functioning of national economies.

In the ordoliberal tradition, politics has the task of creating a regulatory framework in which companies and consumers can operate freely. This includes the idea that 'the prices tell the ecological truth'. The economic costs of a product must be reflected in the pricing. Resource taxes and emission levies are more effective than a large number of individual regulations. Strong cartel authorities should safeguard healthy competition. Political regulation should not replace but should rather enable the free play of forces. There can be no ecological transformation without innovative companies.

If the Greens want to be a party of freedom, they cannot restrict themselves to civil rights, the chocolate side of liberalism. Their environmental, economic and social policies must also meet liberal requirements. The question of whether we see the future as open and flexible or as an era of grim scarcity will be of great importance in determining whether green politics are libertarian or authoritarian. Nature is not the highest instance that tells us how we should live. The better we understand the biological and physical world, the wider the horizon for creative co-evolution of man and nature.

The green revolution

Forty years after the storm that the Club of Rome provoked with its report on the *Limits to Growth*, we are witnessing a renaissance in the growth critique. The call for 'prosperity without growth' appeals to all sides of the political spectrum, from left to conservative. It is prompted by a number of motives: unease regarding the consumer society, the increasingly rapid pace of life, the pressure of global competition. Added to this is the worry regarding the fundamental ecological conditions for life: climate, soil fertility, oceans, air quality and biodiversity.

However understandable these motives are, the fundamental opposition to economic growth is escapist. It ignores the fact that we are confronted by a historic growth period, driven by the rise in the world population and the ambitions of billions of people for a better life for themselves and their children. And it is bogged down in the quantitative logic of 'more or less', instead of facing up to the central question of the *quality* of production and consumption. It is not just the size of the GDP that is the decisive climate factor but rather the way we generate energy, produce industrial goods, farm the land, build cities and organize transport. It is not the standard of living that has to be reduced, but the emission of greenhouse gases, the squandering of valuable resources and the vast quantities of waste of all kinds.

Zero growth is not a realistic option, especially for the large part of humanity that is just entering the industrial modern era. It also applies in the final analysis to 'old Europe'. Most people on our continent do not live a life of ease. Even in a prosperous country like Germany, most labourers, office workers and pensioners just about make ends meet. The further south or east we go in Europe, the larger the number of people living in modest circumstances. They do not have surplus wealth; they lack income, have fewer education possibilities and job prospects and a lower quality of life. This applies even more so to the vast majority of the population of Asia, Latin America and Africa. There are almost a billion people living barely at the subsistence level. Of course, economic growth on its own is not a patent recipe for improving the situation. Whether the poor will benefit from a higher GDP depends on a number of factors – land rights, unions, political participation and investments in education, health and infrastructure. But without economic growth, billions of people on our planet will have no chance of improving their social situation.

Global economic output will double in the next twenty years. This is reassuring and alarming at the same time. It is reassuring because economic growth is associated with lower child mortality, higher life expectancy, better education and rising income. It is alarming because doubling the consumption of natural resources and emissions would lead to an ecological disaster beyond all expectations. The old resource-guzzling and energy-intensive growth model has reached its capacity.

If a 'carry on' mentality would lead to a global disaster and a call for the 'end of growth' would not work, what alternative is there? A central challenge in the next few decades will be the ecological transformation of capitalism, the transition to a method of production giving prosperity to a growing world population without ruining the natural basis of existence. The core principle of this green industrial revolution is the decoupling of economic value creation from the consumption of natural resource. The categorical assertion that this is not possible is as plausible as the assertion of old that 'man cannot fly'.

Growing prosperity for billions of people while reducing environmental consumption calls for the extensive decarbonization of the economy, in other words the gradual phasing out of coal, oil and gas in favour of renewable energies. A second innovative field is increasing resource efficiency. The challenge is to 'make more from less', in other words to create more prosperity with fewer natural resources and less water and energy. A third qualitative change would be the transition to linked material cycles in which all residues are used as the starting point for new production processes. In future only those materials that can be completely biologically or industrially recycled should be used.

This is not a green fantasy. In view of the many alarms being sounded about record temperatures, melting icebergs and islands of plastic in the oceans, it is easy to overlook the signs of a change in ecological thinking. The European Union recorded economic growth of 46 per cent between 1990 and 2014. In the same period, carbon dioxide emissions were reduced by 23 per cent. In Germany, this decoupling is even more manifest, in spite of high export surpluses – it produces more than it consumes. In China, emissions in the energy sector declined in 2014 and 2015 – the result of increased energy efficiency and the rapid development of renewable energies. This is far from being sufficient to sound the all-clear. But it shows that a change of attitude is possible if the right course is set.

Human civilization is reliant on a more or less stable climate, the fertility of agricultural land and intact water cycles. If we overload the ecosystems, serious crises could result. To that extent, there are indeed ecological limits to growth. The important point is that no fixed limits for production and consumption should be inferred from these red lines. The possibilities for people on our planet will not be determined primarily by geophysical factors. Our most important resource is *creativity*. This includes the ability to overcome shortage crises through innovation. The energy factor is also not limited. Solar power is an almost inexhaustible source of energy. This is not just a question of obtaining power and heat from solar energy but of the technical imitation of photosynthesis – the transformation of sunlight, water and carbon dioxide into biochemical energy.

The ecological transformation of capitalism is a huge innovation and investment programme. It calls for a renewal of the industrial apparatus and public infrastructure, a radical change in energy generation and transport, a building revolution and a different form of agriculture. In a shrinking economy, investment and the rate of innovation also shrink. The race with the climate crisis calls for an increase in the speed of structural change, however. This requires *more* investment and a *higher* rate of innovation – in other words, the opposite of a post-growth strategy.

In the wake of the financial crisis in 2007, youth unemployment climbed to record heights in Europe, and many people were impoverished. To enable everyone to live in accordance with their means, there is a need not only for a fairer distribution of wealth but also for a prospering economy with a strong industrial basis. The way out of the crisis is through an innovation offensive to make Europe a leader in ecological modernization. Our continent has the scientific and industrial potential to be at the forefront of the green industrial revolution. The 'energy revolution' is a reference project in this regard. With it, we can demonstrate that the phasing out of fossil energy can be a successful economic model. Only then will it also become a model for the emerging societies in the South.

Limits of denial: the Greens and flying

Reliance on inventiveness and entrepreneurship as productive forces does not relieve us of personal responsibility. It is all very well to eat less meat, to cycle and travel by train, not to buy products that exploit people or destroy rainforests. Everyone is at liberty to seek the 'good life' in more leisure and social relations instead of higher income and consumption. But a sober look at the extent of the ecological challenge shows that the problem will not be solved with an appeal to frugality. 'Less of the same' is not enough. Without a green industrial revolution, we will not win the race with climate change.

There is no clearer indication of the dilemma in the policy of ecological restraint than in aviation. In the environmental scene, flying is regarded as the ultimate ecological sin. Depending on the calculation method used, 5–8 per cent of global greenhouse gas emissions are attributable to air traffic, and the proportion is increasing. Added to this is the amount of land required for new airports and the noise pollution when taking off and landing. Something therefore needs to be done. If we absolutely need a car, we can switch to electric cars or car-sharing systems. There is no choice when flying. Either we fly, or we don't fly. It would seem logical to call for restraint on air travel. There are enough appeals of this type but they have had next to no effect. Worldwide air travel is growing at an annual rate of 4–5 per cent.

Until recently, flying was the privilege of the economic and political elites. The vast majority of passengers came from North America and Europe. This has now changed fundamentally. Air travel is becoming cheaper and cheaper. More and more people can afford to fly, and more and more passengers come from Asia, the Middle East and Latin America. Today, there are around 3.3 billion (!) flights per year. This figure and the number of aircraft are both forecast to double in the next twenty years. More than half of all new aircraft are likely to be purchased by Asian airline companies. At the same time, new mega-airports will be built with capacities far in excess of Frankfurt Airport, for example. It sounds futile to counter this global growth dynamic with appeals to cycle or travel by train instead of flying to holiday destinations.

Flying has in fact been of great benefit to civilization, even if members of the educated middle class turn up their noses at the mass of package-holiday tourists on Mediterranean beaches. Air travel is the analogue worldwide web. It connects people from all over the world, widens our horizons – quite literally – and brings far-off continents closer. Without aviation, there would be no globally linked economy and science, no international cultural exchange and no global civil society. More than any other means of transport, the aeroplane is the vehicle of globalization. International conferences, travel diplomacy, school exchanges, semesters abroad for students and major sporting events all rely on flying. Even love is global today. The number of transnational families is growing. Worldwide migration also causes more air traffic – we need only look at the increase in flights between Germany and Turkey.

In short, flying is an integral component of the modern world. It is somehow schizophrenic for the Greens to fulminate so loudly against aviation, as no other social group flies more often than the green electorate. According to a study by the Wahlen research group in 2014, 32 per cent of SPD supporters and 36 per cent of CDU/CSU sympathizers stated that they had flown in the previous year. Among the Greens, the figure rose to 49 per cent. This is not surprising. They have above-average education and high earnings; they work in international environments and are curious about the world. Apart from heroic individuals who systematically refuse to fly, all appeals to restrict air travel function only to assuage our own bad conscience.

What, then, is the solution to this dilemma? Exactly what the fundamental environmentalists condemn as eyewash: the great leap towards environmentally sound flying. We cannot exempt aviation from its obligation to drastically reduce greenhouse gas emissions by mid-century. But air traffic is likely to grow considerably at the same time. This conflict can be resolved through the development of climate-neutral aircraft. This might sound like science fiction, but researchers are already working on the realization of this vision today. Manufacturers and airports have set themselves the target of putting a cap on their emissions from 2020 and reducing them by 75 per cent by 2050. The first intermediate goal will be achieved through a further reduction of fuel consumption per kilometre and a system of offsets (compensation projects). This is little more than a continuation of the present situation. Thanks to lighter aircraft and more efficient engines, fuel consumption per passenger kilometre has already been reduced since 1990 by a good 40 per cent. With the latest generation of aircraft, a further 15–20 per cent reduction is possible. But the efficiency gains to date have been eaten up by increased air traffic – a classic example of the famous rebound effect.

This trap can be avoided only by means of a technical revolution in aeronautic engineering: better aerodynamics, super-lightweight and yet stable materials, new propulsion technology and alternative fuels. The aircraft of the future will fly with a combination of battery power and regenerative fuels. It will use solar cells and horizontal wind generators to recharge during flight. The competitive market will decide whether biofuel from algae, synthetic hydrogen or methane obtained from surplus regenerative power will play a role.

Politicians should not rely on a single specific technological approach but should set industry clear reduction targets and couple carbon dioxide emission in air traffic with staggered price increases. It is clear that the national regulation of air traffic can have only limited effect in a global market. We therefore require international agreements to reduce carbon dioxide emissions in air traffic. The likelihood of achieving this has become greater with the Paris climate agreement. In the long term, air traffic will not be able to escape the commitments contained in it.

The magic triangle: technology, culture, politics

It is clear that the rapid rate of technological innovation does not exempt us from the need to rethink our ideas about the 'good life'. Technological and cultural changes are two sides of the same coin. On closer inspection, the new lifestyle trends among young intellectuals – no-car mobility, vegetarianism, fair trade, shedding of superfluous possessions, leisure time as the new luxury, reconciliation of work and family – are not so much a new culture of restraint as a form of *considered hedonism*. It aims to reconcile competing values: enjoyment and conscience, professional ambitions and social attachments, consumption and sustainability. The alternative to 'carry on' is 'different and better'.

The ecological transformation of capitalism is not a case of faith healing. It is possible to discover signs of change everywhere. In spite of uninhibited wealth creation, in spite of all the bank scandals and fraud cases, there is a new, more moral approach to the economy. Public discussion of humane working conditions in supplier companies, of environmental destruction when extracting natural resources and of tax evasion by major companies is leading to a new regulation of an economy that has got out of hand. A cultural transformation is taking place in many companies. Young talents inquire not only about salary and promotion prospects but about the meaning of their work. The 'moral capital' of companies is becoming a more important factor in economic success. The avoidance of scandal through the observance of minimum social and environmental standards is just as much part of a sensible business management strategy as efficient resource management. Sustainable investment funds are on the rise. At the same time, we are observing a renaissance of a non-profit economy. Communal municipal utilities, non-profit companies and open-source projects are becoming popular, exchange portals are flourishing and the major automobile companies are developing their own car-sharing systems.

Will that solve the problems? Probably not – and certainly not on its own. Markets, competition and entrepreneurship are essential as innovative search engines. To bring more sustainability to the market economy, however, a political and legal regulatory framework must be provided. The progressive increase in the cost of resource utilization, effective carbon dioxide emission trade, ambitious efficiency standards, a commitment by manufacturers to take back old appliances and vehicles, and an ecologically oriented research and technology policy will all contribute in that regard. But the state cannot manage on its own. It requires innovative companies, a critical public, informed consumers and strong civil society actors who sound the alarm and exert pressure. A combination of these factors would make it possible to overcome the ecological crisis without restricting freedom.

12

Civilizing Capitalism

In public discourse, democracy and capitalism are usually described as opposites: democracy is leveraged by the inherent dynamism of the market; governments and parliaments are put under pressure by multinational companies. It would seem sensible to extend the 'primacy of politics' to the economic sphere: the more state-owned property and the more political control of investments, production and consumption the better. This sidesteps the question of the connection between *political freedom* and a *liberal economy* that leaves as much scope as possible for decision making to businesses and consumers. Just because we reject the radical market dogma 'the less state the better', it doesn't mean that the opposite is true. How do capitalism and democracy interact?

Capitalist production is more dynamic, flexible and diverse than a notion of 'capitalism' that presents this economic system as being immutable would suggest. Its basic features are quickly listed. Those who have studied Marx will be able to repeat them in their sleep: private ownership of the means of production;

wage labour; competition; the continuous utilization of capital as a stimulus for technical innovation and economic expansion. But what do we know about capitalism when we know that? Not much. Its actual shape, its character and its functioning depend importantly on the political, social and cultural context in which capitalist production is embedded. The faces of capitalism are as different as the conditions in which it operates.

The capitalism that Marx and Engels described from empirical observation is completely different from the Western European, American, Japanese or Chinese capitalism of today. In fact we are dealing with multifarious forms of capitalism. Their welfare-state, liberal-market, democratic or authoritarian features differ considerably in terms of the legal and social situation of the working population, the distribution of wealth or the state contribution to the GDP, which varies from almost 30 per cent to more than 70 per cent. Because capitalism is so diverse, politics has an important influence on the functioning of the economy and the life of the population.

Culture (in Marxist diction, the 'ideological superstructure') also influences the economy. This includes religious norms such as the Islamic ban on interest or the Calvinist doctrine of justification, which gives entrepreneurship its theological legitimation, as well as the prevailing political values of a society. Where social equality has high status, capitalism will be more aligned with the idea of a welfare state than in societies with an individualistic freedom ethic.

The capitalist method of production is so successful because it permanently renews itself. If the political and social conditions change, capitalism changes as well. New technologies – from the steam engine to the internet – also produce new forms of capitalism. The decentralized interaction mediated by the market between producers and consumers is a perfect copy of the principle of evolution. It is for that reason that, contrary to all prophecies of doom, capitalism has not been brought to its knees by various crises but has been strengthened by them. It is an intelligent system that transforms crises into innovations. Capitalism also gets new stimulus from opposing social systems: the struggle of the labour movement against Manchester capitalism led to the welfare state, the anti-authoritarian revolution of 1968 encouraged a process of cultural modernization, the fight for gender equality revealed women as a new source of talent for companies. There is a lot to be said for the ecology movement as an innovative driving force behind the formation of 'green capitalism'.

Property for all!

The adaptability of capitalism is also due to its compatibility with quite different forms of government. We find various combinations of capitalist economy and absolutist, fascist, state-capitalist and democratic forms of government. In the longer term, the capitalist method of production can only flourish, however, if there is a minimum of legal security, particularly with regard to private property and freedom of contract. Authoritarian regimes can be compatible with entrepreneurship and competition, at least for a certain time, as long as they limit arbitrary access to private property and have long-term policies.

In Europe, the rise of capitalism and the development of the democratic rule of law went hand in hand, from the Italian city republics and Hanseatic cities to the parliamentary republic. Germany and Italy, like Japan, were special cases as 'tardy nations'. In these countries, large sections of the population made their peace with democracy only after defeat in the Second World War.

There is something to be said for the thesis that capitalism is not reliant on democracy but that democracy cannot exist without a minimum of economic freedom. In the Anglo-Saxon tradition, there is a clear understanding of the interdependence of property and freedom. The democratic republic in America developed as a society of free property owners who were not satisfied with being *bourgeois* but wanted to be *citoyens*, free citizens in a political community whose government had no absolute power over society. In Germany, by contrast, there is not such wide acceptance for the idea that property gives a certain independence and forms the economic basis for civil identity and responsibility.

Abolition of private property! That was the battle cry of revolutionary socialism. It was intended to put an end to all exploitation and to create a society of equals. In reality, the abolition of private property led to a total subjugation of society to the absolute authority of the state and its functionaries. When political and economic power is united in a state, there is no room any longer for the autonomy of the individual or civil society. Democracy is strengthened not by abolishing private ownership of the means of production but by extending it to everyone. The slogan should be 'property for all!' It is a question of participation in the productive capital of society through asset building by the workforce.

There are a number of instruments for this, from participation in company profits in the form of employee shares to capital wage policy, in which part of the pay rises negotiated by the unions and employers is paid into a non-company investment fund. If the unions had employed this model in the 1970s, at a time when there were still significant wage increases, the workers would today have been by far the largest owners of capital in Germany. If employees are co-owners, it is not only their sources of income that are enlarged. They also enlarge their possibilities for deciding on investments and working conditions. This would be a decisive step towards 'stakeholder capitalism', in which workers and salaried staff are also co-owners of private productive assets. Company boards and managers would then act on behalf of the employees and would be controlled by them.

Employee participation in productive capital is also a smart answer to globalization. It strengthens companies' commitment to their location. It makes them more independent of outside financing on the capital market and works against the fixation on short-term profit maximization. At the same time, it also prevents unions from pursuing a short-term wage policy at the expense of the company's competitiveness. It thus furthers the idea and practical implementation of social partnership – the very reason it is contested by the traditional left.

Values and value creation

Capital without democracy, in other words without a system of checks and balances, is a danger to man and nature. There can be no doubt about this. The interesting question is whether a convergence of values and value creation is developing before our eyes, a new variety of sustainable capitalism, in which morality and profit are no longer irreconcilable opposites. This new capitalism is just now evolving and is still mixed up with the old form. Unscrupulous profiteering, defrauding customers and business partners, predatory exploitation of nature, inhumane working conditions and starvation wages are still widespread. But they are no longer a business model for long-term economic success. This applies at least to countries with an enlightened middle class, a critical public, strong consumer associations and a lively NGO scene.

We are used to looking at business in terms of scandals. The pollution manipulation by automobile companies, corruption, tax avoidance artists like Apple or factory farming abuses make the headlines. In the eyes of the public, capitalism is mainly associated with greed and ruthlessness. This distorted picture obscures the changes in the direction of a moral economy. Sustainable business models are based on fairness to employees, customers and business partners and on a responsible approach to natural resources. The avant-garde of this new capitalism is a new generation of start-ups that are not necessarily after the big money but understand entrepreneurship as a way of improving the world. For them, ecological sustainability and social commitment are central driving forces. They can be found in the restaurant and fashion sectors, in e-business and agriculture, in the energy industry and the architecture scene. They are pioneers of a much broader development that is also embracing the 'old economy'.

Under the entry 'sustainable business', the Google search engine lists more than 10 million hits. 'Sustainable investment' has 2.66 million. Both contain countless companies, associations, financial service providers, environmental, international law and human rights organizations, foundations, publishing companies and education institutes. This is an indication of a widespread trend driven by a diversity of actors. The number of companies working according to binding standards for environmentally responsible management and resource efficiency is steadily growing. In 2013, there were over 260,000 companies worldwide certified according to ISO 14001. Large customers generally insist on this certification before placing supply orders.

Earning moderate profits with a good conscience is the guiding principle of a new generation of investors. They are not only idealists. In the long term, ethical investments are more crisis-proof than conventional investment strategies. According to a study by the European Sustainable Investment Fund (Eurosif), sustainable investments[1] in Europe between 2008 and 2012, in other words in the middle of the financial crisis, grew from 2.7 to 6.8 billion euros. The lion's share was held by institutional investors such as insurance companies, foundations, pension funds and church institutions.

At the same time, ethical investment is also the fastest growing sector in public funds. The finance markets are reorienting themselves. Climate protection has become an economic factor. Companies are forced by major investors to reveal their carbon dioxide balance. Stock exchanges are seeing supposedly unshakeable giants in the energy sector being re-evaluated on the basis of their carbon dioxide intensity. Shares in coal companies have taken a plunge, and oil and gas supplies have been strongly devalued. The price erosion for fossil fuels is one reason for this, as is the realization that around two-thirds of known fossil fuels will have to stay in the ground if climate change is to be kept within anything like controllable limits.

A pioneer in the field of sustainable business management is Otto-Versand, the largest mail order company in Europe with sales of a good 12 billion euros. The company produces a detailed annual environment and social report. Environmental management systems aimed at the continuous optimization of purchasing, transport and packaging are in place in all locations in Germany and abroad. Obviously, even a company like Otto has to operate profitably. This limits its scope on the cost side as long as customers are not willing to pay more for high environmental and social standards. But with low profit margins in particular, efficient management of natural resources, water and energy, the avoidance of waste and the optimization of transport chains are effective ways of reducing costs. High safety standards lessen the risk of production stoppages and claims for damages, not to mention the fact that a positive environmental and social image is a competitive advantage that should not be underestimated. Companies such as Puma, Adidas or Nike, which until recently were targeted because they used Asian sweatshops with brutal working conditions, have now introduced minimum social standards for their suppliers. This does not exclude the possibility of scandals occurring at the far end of the production chain, but it essentially improves working and environmental conditions in low-wage countries.

Finally, the human factor is playing a greater role in an economy whose success depends more and more on the knowledge, personal initiative and teamwork capability of workers. The more purely mechanical tasks are replaced by complex development, control and service tasks, the more important human capital becomes in value creation. Attracting and keeping skilled workers is turning into a decisive factor in the knowledge economy. This trend will become even more pronounced with the demographic changes currently taking place.

The shorter the supply of skilled workers on the labour market of the future, the more companies will have to offer their employees in terms of working conditions and development possibilities, and the more they will have to exploit the talent potential presented by women and migrants. The promotion of women, family-oriented working hours, advanced training programmes, profit sharing and diversity management are therefore not luxuries but economic necessities. Successful companies and sites are notable for their cultural and ethnic diversity. It is hard to find a greater mixture of people of different origins, skin colours and sexual orientations than in Silicon Valley, New York or London. Open societies are good for immigrants provided the latter have the opportunity of advancing socially through education and work instead of being kept away from the labour market for fear of competition.

It is easy to argue that all of these progressive developments are marginal phenomena or merely a cosmetic cover for the brutal reality of capitalism. It is always good to measure statements on the basis of action. But left-wing fatalism, in love with failure, is blind to anything new growing out of the old. Those who insist that capitalism cannot be reformed are missing the opportunity for sharing in the shaping of a better world.

Driving forces behind the new

What is behind this most recent metamorphosis of capitalism? The new generation of managers cannot have been untouched by the green wave of the last twenty-five years. The decisive factor is the inclusion of social and environmental targets in the self-interest of modern companies. At least in Europe and other highly developed markets, it is no longer possible for companies to increase profits by ruthlessly exploiting man and nature. Their interest in sustainable long-term added value has caused them to rethink.

In an era of worldwide production and transport chains and a global public, the avoidance of disruptions, production stoppages and scandals are in companies' own interests. Efficient environmental management reduces the consumption of natural resources and energy and avoids waste. Eco-efficiency gives a competitive advantage, all the more so when customers' environmental awareness is high and environmental regulations become stricter against the background of climate change. The time is coming to an end when companies can maximize their profits by passing on ecological costs to the public.

The significance of *reputation capital* for economic success is also growing. Global brands such as Adidas, Nestlé or Shell are vulnerable to scandals that can impact on their market shares and profits. The takeover of Monsanto by Bayer has a lot to do with the company's wretched image. Companies working in the health and food sectors are particularly susceptible to scandal. Negative headlines can jeopardize sales and profit. No company can afford to ignore the growing health consciousness of young people and the educated middle class.

The good news is that ecology and economy are not irreconcilable, and the social market economy is not outmoded. Does that mean that we can sit back and wait for the new sustainable capitalism to gain the upper hand over the old exploitative economy? Not at all. The progress made to date to counter air and water pollution and with regard to recycling and alternative energies has been hard won. Companies generally defend their short-term interests. Attempts to tighten emission limits are met by a fanfare of cries announcing the decline of Germany as an industrial location. In fact, the opposite is true: Germany's pioneering role in many areas of environmental politics has effectively increased the competitiveness of German industry. German companies are among the leaders when it comes to energy and resource efficiency and environmentally friendly technologies and processes. The German share in the export of environmental technology is twice as high as its share in world trade as a whole. This is not trivial: nowhere is the global demand growing so strongly as with 'green' technologies and products.

What I am describing here is a supposition. It is by no means certain that it will be realized. The design of sustainable capitalism based on a responsible approach to man and nature is an optimistic scenario of what is *possible*. The link between capitalism and ecology is a challenge for the twenty-first century. Those who take this path will hold the best cards, both economically and politically.

Politics and the economy

As the tyranny of the strong over the weak is an inherent feature of any limitless freedom, limitless capitalism also leads to crass forms of exploitation. They were typical of the early stage of capitalism, from which Marx obtained the empirical material for his critique: child labour, ruined health, social misery, political disenfranchisement. This ruthless form of capitalism exists today only in those world regions in the early stages of capitalist industrialization. The Marxist answer to such appalling conditions was socialism and revolution. From our experience of socialist revolutions since 1917, however, we know well enough that they are not the way to liberty, equality and fraternity. On the contrary, the concentration of political and economic power at the head of the socialist regime merely breeds new monsters.

The liberal democratic answer to the excesses of limitless capitalism says: 'We must construct *social institutions*, enforced by the power of the state, for the protection of the economically weak from the economically strong.'[2] Popper explicitly rejects the equating of liberalism with laissez-faire capitalism: 'If we wish freedom to be safeguarded, we must demand that the policy of unlimited economic freedom be replaced by the planned economic intervention by the state. We must demand that *laissez-faire capitalism* give way to an *economic interventionism*.' At the same time, however, state intervention in the economic process should not get out of hand. It is justified only to the degree that it is required to safeguard the economic freedom of all.

In spite of the short period of neo-liberal deregulation at the end of the twentieth century, there has never been such tight regulation of private property. Anyone starting a company has to deal with thousands of regulations: labour law, environmental law, social security statutes, building codes, consumer protection, liability law, anti-discrimination laws, worker participation, tax law, and so on. The number of relevant laws and regulations has increased fourfold in the last few decades. Scandal-loving media, unions, consumer associations and environmental organizations ensure the constant addition of new norms. The bureaucracy grows with it. Those who ask why the economic dynamic in western societies is slackening can find at least some of the answer there.

Against utopianism

Popper justifies the advantages of minimally invasive intervention as an answer to the authoritarian fantasy of comprehensive planning and control. It has fewer unforeseen side effects and strengthens rather than replaces the responsibility of citizens and companies. It should also be reversible if it does not prove itself. Messianic politics aims at maximizing state intervention, while liberal politics is restricted to the minimum. It does not claim the absolute truth or promise heaven on earth or the end of all crises and conflicts but concentrates on the gradual improvement of conditions in the here and now. It does not follow any divine purpose but feels its way on open terrain. It learns from mistakes and is prepared to correct itself.

Popper calls this type of politics 'piecemeal technology' or 'social technology in small steps'. This sounds more technocratic than it is meant to be. Popper does not plead for supposedly apolitical rule by experts, nor does he turn politics into a form of applied engineering. The competition between rival political ideas is a central component of democracy. The rightness or wrongness of an idea is only revealed in practice, however. The superiority of the 'open society' is demonstrated, apart from anything else, in its *willingness to accept errors*: as every political action carries the risk of error, the mechanisms for criticism and correction are of decisive importance. These include the separation of powers, party pluralism, the possibility of voting out governments, the interplay between the government majority and the parliamentary opposition, independent media and an active civil society. They are the reasons for the greater learning capacity of democratic governments compared with authoritarian ones.

Popper's scepticism with regard to all types of political utopianism is empirical. Its political component evolved against the background of the totalitarian rule that engulfed a large part of Europe in the late 1930s. Popper sees an inherent connection between utopianism and totalitarianism: 'The attempt to make heaven on earth invariably produces hell.' It opens the doors to 'intolerance [. . .] to religious wars, and to the saving of souls through the Inquisition'.[3] There is no more dangerous political idea than 'making others happy' because it tempts us to impose our scale of values and to intrude on their privacy. Popper's answer is a simple political form that offers specific answers to specific challenges without promising lasting solutions. Modern societies are extremely complex and fragile structures. Everything is mutually dependent and every intervention in the social structure triggers a number of interactions that we cannot completely foresee. Well-intentioned measures can ultimately have more negative than positive effects. For that reason as well, minimalistic interventions on the principle of 'only as much as is absolutely necessary' are generally more useful than large-scale social experiments that carry the risk of failing disastrously.

Transforming a famous slogan by Rosa Luxemburg, the true alternative for our time is not socialism or barbarity but *democracy or barbarity*. We have no other safeguard than liberal democracy to ensure against a return to nationalism and the persecution of those who think differently.

Institutions as a counterweight to the power of capital

In Popper's opinion, indirect control – through the legal framework, taxes and levies – is preferable to direct intervention by the state in economic decisions. In claiming so, he anticipates the guiding principle of ordoliberalism: the state should create a regulatory framework for economic actors but should not replace them. This means creating *institutions* as a counterweight to the economic power of capital to assert public interests: unions and cooperatives, universal suffrage, social insurance, cartel authorities, free press, and so on. This is precisely what has gradually happened since the wild era of Manchester capitalism – as a result of hard-fought battles and the activities of social reform parties and trade unions.

Marx would hardly recognize present-day capitalism, at least in its European version. The life expectancy of workers has doubled. Political freedoms, education, health care, leisure and the quality of housing have reached a level that the utopian socialists could hardly have imagined. Universal suffrage forces the government to take account of the interests of the broad mass of voters. A comprehensive set of laws to protect workers, consumers and nature has been gradually elaborated. The fact that European authorities can impose fines of billions of euros for cartel agreements is an achievement of democracy, as is the progressive taxation of income and security against the risk of unemployment, illness and care dependency. None of this has solved the 'social question' or removed economic inequalities. The error is to assume that these questions have to be resolved once and for all, instead of regarding them as ongoing tasks that need to be repeatedly addressed anew.

13

Shaping Globalization

Is it over-optimistic to claim that democracy has civilized capitalism? It might be argued that this applies only to the democratic societies of the West and even there not in every case. Moreover, with the boost that capitalist globalization received in 1989/90, when the 'socialist world system' collapsed, the balance of power between politics and the economy has shifted again in favour of capitalism. While democracy remains essentially linked to the nation-state, capital and goods markets have long overcome these limitations. Banks, investment societies, industrial companies and internet businesses operate worldwide, shifting huge amounts in seconds, playing off states against one another and escaping national regulation. It is not governments controlling companies but international capital that dictates its conditions to politicians. This narrative, originally launched by a left critical of globalization, is now readily taken up by national populists, who are setting themselves up at the forefront of opposition to globalization. In the name of national sovereignty, they want to put customs barriers in place again and restrict the free movement of capital, goods and people. It is a widely and firmly held view that globalization undermines democracy and that politics is merely the agent of worldwide capital. Whether deliberately or not, the worshippers of this thesis are fuelling the preconceived idea of the powerlessness of elected governments and are feeding the frustration with the political system.

Let us take a closer look. First, it is not true that the social containment of capitalism is restricted to Europe and its offshoots. Japan and South Korea have produced their own models, and a number of countries in transition have more or less stringent work protection laws, regulations for maternity leave, limited working hours or provisions to prevent child labour. The International Labour Organization has 175 members devoted to the observance of core working norms. These include freedom of association and the right to collective bargaining, the abolition of forced labour and child labour and a ban on discrimination in employment and occupation. The ILO General Assembly has also adopted a number of labour and human rights conventions – for example, to combat the increase in slave-like working conditions, occurring mainly in Asia, but including forced prostitution in Europe as well. It is up to the individual member states to ratify these protective laws.

In many cases, the deficit is not in the legislation itself but in its practical implementation. Either there are no supervisory authorities, or else the state organs are in collusion with businesses. Where there are free unions, an independent judiciary and critical media, violations of protective laws are more likely to be avoided than in authoritarian regimes in which employees are exposed to the arbitrariness of local potentates, state functionaries and unscrupulous entrepreneurs.

Global progress

Second, the widely held thesis, encouraged by populists on both the left and the right of the political spectrum, according to which the globalization of capitalism is leading to a 'race to the bottom', a permanent competition for lower wages and social standards, is also false. In reality, the opposite is true. In spite of the outrageous examples of exploitation that come to light now and again, the worldwide situation of the working class – in terms of legal status, life expectancy and standard of living – is generally improving. According to UNICEF, child mortality is half of what it was in 1990. Since that time, around a billion people have escaped from abject poverty, while the world population has grown by a further billion. Literacy in the poorest developing countries rose from 47 per cent to 73 per cent between 1970 and 2000. Between 1999 and 2011, the number of children with no schooling dropped from 108 to 57 million. At the same time, the number of people with higher education is rising rapidly in developing countries. The global middle class grows annually by 80 to 100 million people.

There are multifarious reasons for this social progress. The main factors are the rising levels of education, economic performance and pro capita income in the emerging economies. International conventions on labour protection and combating poverty also play a positive role. International development aid and private foundations contribute to improved education and health care.

Finally, the combination of a worldwide economic network and a globally linked critical public has helped to improve working conditions and social standards. Modern companies in Europe and the United States can no longer afford to operate without taking the well-being of people and nature into consideration. They are monitored by NGOs and critical media. Child labour, sweatshops without work protection, overlong working hours and environmental damage make scandalous headlines. They damage the reputation of companies and jeopardize their economic success. More and more customers are no longer merely asking how cheap or expensive a product is. They want to know how it was made, whether people were exploited, forests destroyed or animals maltreated. The demand for fair-trade products is growing rapidly. For many consumer goods – including furniture, paper, clothing, coffee, fish, electronic appliances and diamonds – there are certificates calling for a minimum of environmental and social compatibility. At the same time, there is a growing worldwide demand for well-trained and well-motivated workers. Alongside globalization, the levels of education also rise. None of this has put an end to disgraceful inequalities, but it has contributed to improved living conditions for billions of people.

Globalization has not only given a boost to the transnational linking of capital markets, services and production chains. Becoming part of the global market leads to a growing urban middle class in the young industrial nations. It fosters international exchange at all levels. There are hundreds of thousands of Chinese students studying abroad. At the same time, the compression of time and space has produced a globally networked civil society, brought about the globalization of norms and standards and created a global public sphere. Events that used to take place in the semi-obscurity of local or national power constellations have now become the object of global debate, negotiation and regulation.

The internet gives access to global sources of information. Even where the new media are monitored and censored, it is impossible to completely obstruct access to critical information. NGOs play the role of watchdogs over international companies. When rainforests are cut down, a drilling platform springs a leak or a textile factory in Bangladesh burns down, NGOs and the media can expose these scandals overnight. This is also the reason for the reprisals against critical NGOs in states that profit from the ruthless exploitation of their natural resources – including the murder of environmentalists by paid killer commandos in Latin America or South-east Asia.

Is the nation-state finished?

There is no doubt that the controlling power of national politics has diminished as a result of globalization, at least when measured in terms of the traditional yardstick of 'national sovereignty'. The idea that nation-states can regulate all major concerns in splendid isolation no longer applies even to major powers such as the United States, China or Russia. Climate change, terrorism, world trade and financial markets, migration, maritime and air traffic are areas that clearly demand international cooperation, rules and institutions.

Does that mean that the nation-state is finished? Such an assertion would be premature to say the least. The new formation (or restoration) of nation-states from the bankruptcy of collapsing empires did not really start in earnest until after the Second World War. It was given a new boost by the end of the Soviet Union. From Leipzig and Warsaw to Kyiv and Tbilisi, democratic freedom is as much an issue as national sovereignty.

Emerging states such as China, India or Brazil are just now appearing on the world political stage. And in Washington a president has made 'America first' his slogan and is dismantling the multilateral order that the United States was instrumental in creating. The vast majority of nations are not even considering the renunciation of their sovereignty. This also applies to Europe. The Brexit referendum is only the culmination of a widespread trend. There is nothing that idealistic appeals can do about this. On the contrary, the more cosmopolitan the liberal elites, the more stubbornly the majority of the population will hang on to their nationhood.

For most people, the nation-state is still the basic structure for political self-determination, social solidarity and collective identity. This becomes clear above all in times of crisis and danger. Recognition of the continued significance of nation-states is not equivalent to a rejection of supranational cooperation and European integration. On the contrary, the major political challenges can be overcome only through international cooperation, and for the time being the subjects of this cooperation will remain the states.

Global governance

Globalization of the economy is a precursor to globalization of politics. Companies have long been operating beyond national borders. Financial markets are globally linked and transactions of dizzying proportions are concluded in seconds twenty-four hours a day. Companies control their investments as a function of local conditions and market prospects around the world. People have also become increasingly mobile – from the cosmopolitan elites who can choose where they dock to the armies of modern transit workers who cross borders in search of work and pay.

For the first time in history, there is a worldwide market for capital, goods and labour. This shifts the balance of power between national politics and international companies. At the same time, states with attractive domestic markets and high growth potential are still in a strong position compared with international companies. China is the best example of this: instead of allowing itself to be dictated to by others, the Chinese government decides who can do business with China and under what terms – and everyone is queuing up to do so. The United States is also strong enough to impose national rules on the world. Countries on which the American administration imposes sanctions have difficulty in finding international business partners. In Russia, the government fixes the rules by which foreign companies can participate in the exploitation of domestic oil and gas reserves. The EU imposes billion-euro fines on internet giants such as Google or Amazon to force them to adapt to European competition and data protection regulations.

The states in the world that have neither strategic natural resources nor a strong competitive position are in a much worse position. They can be manipulated by financially powerful companies to obtain preferential tax rates, negotiate subsidies or gain exemption from national legislation. The poorer the country and the more corrupt its authorities, the better this practice functions. The shifting of profits by multinational companies to tax havens to avoid taxation is conditional on the existence of states that offer tax avoidance as a business model. Binding multilateral agreements are required to prevent such dumping practices: minimum taxation on corporate profits, information exchange between tax authorities, fixing of minimum social and environmental standards. It helps if economically powerful states join forces to pressurize recalcitrant ones. And it would be at least a first step if the EU were to do its homework and finally agree on a minimum corporate taxation rate.

In spite of the asymmetry between the global economy and national politics, there is no question of globalization being uncontrolled. In the last few decades, the network of international institutions, treaties and regulations has become increasingly dense. At the centre of the global governance system is the United Nations with its numerous subsidiary bodies, including the International Labour Organization (ILO), UNESCO, the World Health Organization (WHO), the United Nations Conference on Trade and Development (UNCTAD) and the World Intellectual Property Organization (WIPO). A second group is responsible for global economic governance: the International Monetary Fund (IMF), the World Trade Organization (WTO) and the World Bank. A third form of coordination is the club organizations such as the Group of 20 (G20), in which the most influential industrial and developing countries reach joint agreements. Among the global environmental policy actors are the United Nations Environment Programme (UNEP), the Montreal Protocol on Substances that Deplete the Ozone Layer, the Convention on Biological Diversity and the United Nations Framework Convention on Climate Change as the summit of global climate diplomacy. Added to this are specific regulations such as the Antarctic Treaty System designed to protect the sensitive ecosystem in the South Pole region from economic exploitation. The list could go on for pages. The global level is supplemented by regional organizations such as the European Union, the Organisation of African Unity or ASEAN. National governments set the tone in all of these bodies. They no longer have the floor to themselves, however, but share the international space with non-state actors: associations of towns, scientific and cultural associations and a worldwide capillary system of NGOs ever ready to put pressure on companies or governments.

Together they form a system of global regulation and control – global governance, in other words – linking state, supranational and civil society actors. Economic globalization therefore fosters the formation of a global civil society and the development of global cooperation. Recent events, however, have shown that this trend is not irreversible. We are in danger of reverting to a political approach that sees world trade and world politics as a zero sum game: a gain for one side is a loss for the other. The withdrawal of the United States from the Paris climate agreement and from multilateral trade agreements sends out alarm signals. Globalization needs to be shaped. To roll it back would be reactionary.

14

How We Can Relaunch the European Union

However piebald the diverse nationalist parties in Europe, they all share a rejection of the European Union. For them, it is an abomination that goes beyond the loose cooperation of sovereign states to form a 'Europe of fatherlands'. According to their propaganda, the EU is an undemocratic power whose intention it is to keep on imposing new and senseless regulations on citizens, a bureaucratic monster, a Trojan horse of globalization and a money-wasting machine at the expense of ordinary people. This is complete nonsense, a cruel distortion of the reality of Europe. The only worrying fact is that this type of 'post-truth' propaganda lands on such fertile ground. The British Brexit vote was not an exception. The supporters of European integration are on the defensive, and their opponents have the upper hand. How can we reverse this trend?

Advocates of European integration traditionally have three arguments: the EU safeguards peace in Europe; it is the driving force behind growing prosperity; and it makes it possible for Europe to shape globalization according to its own rules. All three reasons for a strong European Union remain valid, but they have all become brittle. The EU is helpless in the face of the wars at its outer borders (Ukraine, Syria). It is seen in over-indebted euro countries not as a guarantor of prosperity but as the agency of a hard austerity policy without consideration for the social consequences. And it has been unable to allay the fear of global competition. The trade agreement with Canada has still not yet been finalized, and similar negotiations with the United States have broken down. Free trade is seen as a threat to European consumer and environmental protection standards and as a danger to the European public service tradition. At the same time, the internal contradictions in the EU are becoming stronger. Differences regarding refugee policy have fostered national egoism and a 'not in my backyard' mentality. The United Kingdom is going its own way. The governments in Poland and Hungary ignore the elementary principles of European democracy, particularly the independence of the judiciary. Populist and other anti-EU forces are gaining ground in other countries as well.

There is an evident need for a new attempt to relaunch the European project. We must agree what the European Union can and should be.

The EU is considerably more than an interstate cooperation. That is its strength and problem at the same time. It is a union of democratic states and also a union of European citizens with a directly elected parliament. European legislation, in which the nation-states and the European Parliament are involved, creates a supranational legal space with its own jurisdiction. This is all much more than a mere alliance of states but also less than the 'united states of Europe'. Based on the criteria of classic statehood, the EU is an incomplete and deficient construct. There is no European government, nor does the European Parliament have all the attributes of a sovereign representative of the people. Professional Europeans in the Commission and Parliament in particular rue the fact that there are no European taxes.

If we change our perspective, away from the idea that European integration must culminate in a European superstate, the EU, with its cooperative decision making and dual character as a union of states and of citizens, is in fact very modern. We can learn from multinational companies that top-down centralization is anachronistic. The trend is towards decentralization, autonomy and flexible cooperation. European nations are too self-willed for them to submit to a centre of power. Above all, however, Europeans should regard their historically evolved cultural and political diversity as a strength rather than sacrificing it to the delusional idea of progressive convergence.

Obviously, a single market also requires common rules and standards, even more so with the euro as the common currency. But these binding norms should be kept as lean as possible. This would be a break from the traditional logic of the EU Commission, intent on continually enlarging the common rule book. The European Union is strong as a result of its self-limitation and not through the excessive regulation of everything and everyone. European community policy should concentrate on the major structural tasks in which the EU can function as a global actor: foreign and security policy, world trade and the regulation of financial markets, climate and energy policy, refugees and migration, strengthening Europe's economic competitiveness and dismantling internal inequalities. This is where it has to deliver. Political scientists call this 'output legitimation': it gains approval through specific answers to specific challenges.

Unity in diversity

The EU will survive only if it follows the guiding principle of 'unity in diversity'. European unity stands and falls with the complex balance between member states and European institutions. Those who attempt to shift it once and for all in one or other direction, setting up a European superstate or concentrating all power again in the nation-states, will simply destroy European cohesion.

We should not be fooled by the populists who peddle the idea that the present decision-making structure in the EU is undemocratic. This is not true. All decisions in Brussels are made by democratically legitimized committees: the European governments, which are subject to control by their parliaments and public opinion, the directly elected European Parliament and, in executive matters, the Commission, which is both authorized and controlled by the Council of the European Union. Then there is the instrument of the European Citizens' Initiative, by which non-parliamentary actors can place their interests on the European agenda.

For supporters of a European superstate, there is only one course: transformation of the European Parliament into a classical legislative body and the Commission into a European government. Whether this would make for greater democracy is more than questionable. The further centralization of political power in Brussels is seen by most of the European public as providing not more but rather fewer possibilities for democratic influence. Rightly so: the further away the decision-making level, the less citizens can influence actors, discussion and processes. This applies especially to a structure in which the majority of the population do not speak the same language. Involvement in a political process demands a minimum of familiarity with the personalities, parties and debates. At the European level, this is something that specialists at best can claim. A 'European people' remains a fiction for the time being. In reality, the EU is an association of European nations with very distinct political cultures and traditions. A transnational political public with European media and debates is slow in developing. For all these reasons, European democracy must maintain a strong footing in national parliaments and municipal authorities.

The EU is an attempt to enable small and medium-sized states to safeguard their scope for action in an era of globalization by bundling their sovereignty. The individual states abandon some of their sovereignty to be able to operate more effectively together than they would be capable of alone. A combined front is the only way to prevail in world politics. The United Kingdom will find this out if the country really leaves the EU. Brexit supporters will experience a reality check if their country tries to go it alone against the genuine major powers. In the interests of Europe, everything must be done to keep ties with the United Kingdom as close as possible. We need the British with their sense of global politics and their liberal tradition, which will hopefully withstand the current confusion. Apart from anything else, there is the question of European security policy, which is difficult to imagine without the United Kingdom.

15

What is at Stake

What is the answer to the crisis of modern liberalism? There is not one *single* answer, as the problems are too complex. But we know the direction we should be looking in. One core point is to restore confidence in the future. Security and a sense of belonging are elementary needs, particularly in turbulent times. These are lessons that can be taken from Brexit and Donald Trump's election victory. To ignore them is to lose. At the same time, they cannot be recovered through withdrawal into nationalism and the fiction of cultural homogeneity. That is an empty promise made by demagogues. In the world of the twenty-first century, the only security is *in change* or, to put it more pointedly, *through change*. This requires self-confident people capable of action and strong republican institutions: a public education system that offers everyone the possibility of developing their talents; social security systems that protect from poverty; and parliaments whose discussions reflect society.

Public support for democracy depends on a number of factors, some material, some immaterial.

Freedom is a strong motive. *Love of freedom* is one of the most important democratic virtues. But on its own it is not enough. One of the guiding principles of an open society is upward social mobility. It is a matter of equal opportunity, the real possibility of social advancement for all. In the last twenty years, our societies have distanced themselves increasingly from this promise. Those who work hard but still do not get anywhere either give up or store their rage, to be activated at a suitable moment.

The vast majority of citizens must feel that burdens and advantages are fairly distributed. However fairness is defined, when the disparity between rich and poor becomes greater, confidence in political institutions suffers, clearing the way for populists on the left and right. In times of turbulent change, the need for security and a sense of belonging grows. Crisis-proof basic social security also assures democracy.

The foundations of democracy are undermined if there is a feeling that politics is merely a game of uncontrollable forces. This applies to the restraining of financial markets and the taxation of transnational companies. It applies as well to the issue of refugees and migration. The question of how many people from far-off countries a state should welcome cannot be decided outside the democratic decision-making process without antagonizing the population.

Confidence in democracy is not just a question of the current political and social situation. At least as important are future expectations: are people optimistic or disheartened about the coming years, do they look forward with assurance or anxiety? The enemies of an open society invoke the fear of loss of control, of social decline and an uncertain future. To pull the carpet from under their feet, we must transform anxiety into optimism: Yes we can! We are capable of building a better future.

Democracy requires political leadership, in times of crisis even more so than when things are going well. Democratic leadership does not mean 'obey me'. It listens and seeks dialogue without telling people what they want to hear. It has the courage to make decisions on the basis of sober reflection. It exposes the pros and cons instead of simply stating that there is no alternative. And it provides orientation in confusing times.

Security in transformation

To play off freedom and security against one another is a cardinal error, both theoretically and politically. Security is one of the fundamental social needs of humanity, or at least of the majority of people in society. It is no coincidence that the 'right to safety' had a prominent place in the Declaration of the Rights of Man and of the Citizen in the French Revolution. Liberal democracy can exist only if it conveys an adequate degree of security. No type of government can guarantee absolute security, in other words protection against physical violence, criminality and social hardship. But a certain feeling of security is an elementary prerequisite for freedom in practice. Fear makes one unfree. This applies to political commitment and to work and to the way we move in public. Do we feel safe or anxious? This is a fundamental question. It depends less on our individual disposition than on the political and social situation.

Conversely, a free society will generally guarantee a higher degree of security. Independent unions, welfare associations and competition for voters' favour guarantee that social issues have a high priority in democratic politics. Parliamentary democracies run a greater risk of overdoing the welfare state by extending welfare benefits at the expense of solid state finances and economic performance. Parties are not doing their job if their election campaigns are run on the basis of 'who offers more'. In doing so, they underestimate the ability of voters to think beyond their direct interests.

Politicians worthy of the name must have the courage to address unpleasant truths and to present the electorate with clear alternatives instead of promising them pie in the sky. Jean-Claude Juncker's *cri de coeur*, 'We heads of government all know what to do, we just don't know how to get re-elected after we've done it,' suggests that voters are only interested in their immediate advantage and just want to receive benefits all of the time. According to this attitude, governments that respect their citizens in any way are committing political suicide. So they prefer to wait for the problems to go away of their own accord. This attitude appears to be confirmed by the fact that European governments implementing austerity measures at the high point of the financial crisis were subsequently voted out of office. It might be useful to look more closely at the reasons, however. Was it the lack of prospects and social imbalance in the austerity policies and the authoritarian way in which they were implemented? Were the ruling parties getting their comeuppance for the long fomenting loss of confidence in establishment politics? Margaret Thatcher was re-elected twice. Gerhard Schröder stumbled above all on account of his 'basta' (enough is enough) style and his inability to provide a narrative for his reform agenda that would be acceptable to the unions and veteran Social Democrats. Those who have respect for their voters must show that they do not trifle with their sense of fairness. This applies not only to Social Democrats.

The fact that democracies must guarantee fundamental security should not result in responsible citizens being transformed into wards of the welfare state. Liberal social policies should rather strengthen the ability of individuals to determine the way they lead their lives. People who feel merely that they are the passive objects of the changes taking place around them are easy prey for demagogues and populists of all hues. It is therefore important to make them *active subjects of change*, capable of dealing confidently with the newness. A central prerequisite for this is education, education, education. The education system should make possible the optimum development of the capabilities of all. At the same time, it is a republican institution that should promote a sense of identity. Since the abolition of compulsory military service, general schools are now the only mandatory institutions that bring together people of different origins and social classes. At university, everyone can go their own way, but not in school. They have to learn to get on with one another. This applies to pupils, teachers and parents alike. Schools are republican institutions, in which democracy can and should be learned, not only as a subject but in practice. In view of the increasing social differentiation and bloc formation, there is a need for places where citizens are treated as equals.

We are the state!

Democracy is not just a system of institutions. It depends on the *republican virtues* of its citizens, the *res publica*. Obedience to laws is a sign of subservience. By contrast, democracy is based, according to the classic formulation by Montesquieu, on 'love for the republic'. While an absolutist community ('you are nothing, the nation is everything') destroys the freedom of the individual, a purely individual concept of freedom ignores the fact that individual freedom is possible only in the context of others. We are citizens only in so far as we are interested in public matters and pursue joint projects with others that correspond to our idea of the common good.

It is for citizens to commit on both a large and a small scale to the just organization of their communal life. Without the unpaid activity of millions, our society would be much poorer and colder, and without the participation of civil society in political life, democracy would be but an empty shell. In contrast to an authoritarian state, a democratic republic is the work of its citizens: 'We are the state!' We, the people, give ourselves a constitution, elect our representatives and participate in many ways in public life.

Interest in and commitment to public life cannot be coerced. It is voluntary. Just as the democratic state cannot dictate how its citizens should live, it cannot force them to take part in political and social life. Otherwise, we would soon find ourselves on the slippery slope of moral crusading. It is nevertheless legitimate to ask how the life elixir of democracy can be maintained. This applies all the more to a migration society composed of people with quite different political socialization.

A minimum of *political education* is essential to be able to live in a modern society. It provides an orientation, strengthens the ability to make judgements and encourages involvement in public affairs. Political education is also the antidote to 'post-truth' propaganda in the social media, the flood of conspiracy theories and opinion manipulation on the internet. The era of political decision making shaped above all through meetings, newspapers and television is coming to an end. The decisive battles are fought today on the internet. Donald Trump has next to no support in the traditional US media. His campaign is aimed at an alternative public, a parallel universe of like-minded people who have created their own forums in social media. Progressive politics must take up the fight for hegemony on the World Wide Web. This calls for the development of new formats and a new communication style combining emotions, information and sharp argumentation.

Excursus: citizens' dividend and civic work

An interesting approach to fostering civic engagement is the linking of a basic *citizens' dividend* with *civic work* for the community. It takes up the demand for a guaranteed basic income but combines it with a self-chosen, socially meaningful activity. While the 'unconditional basic income' is tantamount to a basic pension for all, financed through taxes, paid from cradle to grave without any performance in return, the concept of civic work rewards social commitment. It limits the recipients and hence the costs and is based on the principle of reciprocity: for a social service (tax-financed citizens' dividend), a service in return is expected (self-chosen civic work).

The model could be attractive both for the unemployed who have no prospects in the short term of a return to the labour market and for those with jobs who wish to stop working for a while and do something for a good cause. Every recognized non-profit organization could offer such jobs, which people could apply for depending on their inclination and qualifications, be it in the social, cultural, environmental protection or development aid sectors. They would only have to provide the infrastructure and assume the running costs, while the state would pay the basic salary and insurance contributions for two years. The extent of the dividend should be much higher than the minimum social security benefit (including housing allowance).

Public acceptance of such a model would probably be much higher than the demand for an unconditional basic income and it would cost much less. Everyone would ultimately benefit: those who can fulfil their wish for a career change and a meaningful occupation, the projects they work in and even the companies to which they ultimately return with new experience and new motivation. Social capital (the sum of social activities in the broadest sense) would grow. If the concept were to prove itself in practice, the entitlement to the dividend could be extended to four or six years during a person's working life.

Models such as this might also provide an answer to the increasing automation that is replacing humans by intelligent machines. If value creation is shifted increasingly to intelligent machines, there is more scope for freely chosen activities that could be financed from the proceeds of automated production processes and services. The civic work concept would help to increase the possibility for self-chosen activity without invoking a utopian land of milk and honey. It would increase economic autonomy compared with the tacit necessity for wage labour and would also strengthen the links between the individual and society.

Republican institutions

The challenge of reconciling the rapid economic and social changes in the modern world with the need for identity and security requires that we think and act beyond the framework of the nation-state. We must develop global governance, crisis prevention and management at the supranational level. The gap between the global economy and national politics must be reduced, the network of international cooperation enlarged. Because the current dynamic tends in the direction of re-nationalism and singing the praises of national sovereignty, international security and stability are highly dependent on supranational institutions. They require a permanent negotiation process and offer a framework for reconciling diverging interests. They do not guarantee that international politics will become civilized. International law is ultimately in the hands of states. But as an alternative to a Darwinian landscape in which might is right, the network of global institutions presents a much better prospect for cooperative conflict settlement.

For the EU in particular, its institutions – the Council, Commission, Parliament and European treaties – provide the framework that ensures cohesion in uncertain times. European unity is in a critical phase. It is under pressure from within and without. At the same time, action by Europe is more than ever in demand. The break in the blueprint for lasting European peace under the Russian leadership, the confrontational attitude of Donald Trump and the progress of nationalist, populist and extreme right-wing forces in Europe will hopefully act as a wake-up call for European politics. The common foreign and security policy needs to be promoted. The establishment of a *European Security Council* to coordinate foreign policy and prepare joint decisions would be a sign of assertiveness in stormy times.

At the national level, republican institutions are vital to social cohesion and political stability. Participation by citizens in public affairs is not just a matter of elections, referendums and citizens' initiatives. There is an important network of public institutions that can be shaped in one way or another by citizens. They form the basic framework for a democratic republic. This category includes the education system, theatres, museums and libraries, social insurance with its self-governing bodies, communal utilities and public transport companies, state foundations and public broadcasting.

The neo-liberal zeitgeist was wrong to regard public companies – from the railways to water utilities – as objects for privatization. Cost efficiency is also necessary in the public sector; it is paid for, after all, with taxpayers' money. But public transport companies, energy utilities, kindergartens and clinics are more than just business enterprises. They provide public goods and enlarge the scope for local democracy, in which the urban citizenry are entitled to a say. Even a large organization like the railways should not be reduced to purely business management logic. An efficient and customer-friendly rail system is a republican achievement. According to Allianz pro Schiene, the public sector in Switzerland invests 383 euros annually per citizen in the railway infrastructure, compared with 52 euros in Germany. Even taking into account the more expensive topography of Switzerland, these figures reflect the different value attached to public transport. Budget questions mirror political priorities.

Alongside elections and referendums, public institutions are the constituent elements of the democratic republic. They permit social and cultural participation and make for a democratic public. Universities, theatres, museums, railway stations and town halls used to be a source of civic pride and should become so again. Investments in the sociocultural infrastructure are at the same time the most effective social policy. Good public child care, reliable all-day schools, communal libraries and swimming baths are more important for families than a few additional euros in child allowance. A region-wide and reasonably priced public transport network is of particular benefit to those who cannot afford a car. The fact that it would also tempt car owners to use public transport is a welcome bonus. A modern nationwide broadband network is one of the most effective ways of assisting small and medium-sized enterprises. For that reason, federal, state and municipal budgets should also give top priority to investments in the social and material infrastructure. They would do much more for equal opportunity and social participation than a modest increase in basic social insurance.

This is our land

When Pegida, AfD and the rest style themselves as German patriots, many people feel the need to speak out in opposition to the claims made by these noisy right-wing troublemakers. They sound a new democratic patriotism by contrasting the distorted image of an old-fashioned, xenophobic, culturally homogeneous Germany with an alternative cosmopolitan and tolerant Germany. For many left-wingers, this word 'patriotism' is irretrievably contaminated. For them, it is just repackaging of the old ugly nationalism. And yet there is a long tradition of left-wing patriotism, starting in Europe with the French Revolution. In Germany, it was seen on the barricades in 1848 (the black-red-gold flag and German national anthem recall this). After the disastrous era of Nazism, the SED sought to present itself as the better side of German history. Bertolt Brecht's *Children's Hymn* dates from this time:

> Grace spare not and spare no labour, passion nor intelligence that a decent German nation flourish as do other lands.
> That the people give up flinching at the crimes which we evoke and hold out their hands in friendship as they do to other folk.
> Neither over nor yet under other peoples will we be from the Oder to the Rhineland from the Alps to the North Sea.
> And because we'll make it better, let us guard and love our home, love it as our dearest country as the others love their own.

At a time when we are on the way to a united Europe and living in a multi-ethnic society, is it still possible today to speak seriously of democratic patriotism without showing ourselves up to be hopelessly old-fashioned?

Yes, it is. It would be wrong to believe that nations would dissolve in the 'united states of Europe' like sugar cubes in a glass of water. They are still an essential resource for emotional identity and political self-determination in times of rapid change. The central conflict is not a question of national or post-national but of how we understand the concept of nation: as a political community constructed on the basis of shared values and actions, or as a community of destiny based on common origins and cultural traditions.

'Constitutional patriotism' is the opposite of mystical nationalism. National identity is created not from the dark primordial source of the German soul and way of life or through the fiction of shared origins, but through the political association of free citizens. This does not mean that we should ignore history. Constitutional patriotism also requires grand narratives connecting past and present and explaining the meaning of a liberal constitution. Present-day Germany was shaped by key events in German and European history: the democratic revolutions of 1848 and 1918, Nazism and the extermination of the Jews, the Second World War and the partition of Germany, the peaceful revolution of 1989 and reunification, and the integration of Germany in the European Union and NATO. These stories have to be told repeatedly, reinterpreted and passed on to the young generation and to immigrants, who should be aware that they are coming to a country with a specific history.

The attack by the anti-liberal front aims at turning the clock back to a world without anti-authoritarian education, feminism, gay and lesbian parades, mass immigration and a hedonistic consumer culture. The 1968 student movement is seen as a major lapse, which needs to be rectified. AfD and the others want a different Germany. They are rebelling against a land that over the years has become more tolerant, open, environmental and international – against our land. Democratic left-wingers, liberal free spirits, feminists and green do-gooders are no longer marginal. We have changed this society and are part of it.

The Basic Law was a conscious new beginning, a counter-programme to large parts of German history characterized by nationalism and obedience to authority. At the time, it anticipated the (West) German reality. In fact, the early West Germany was not a clean break, neither in terms of the personalities involved nor ideology. It could not have been so. The constitution was completely rewritten, but the people who built the new state carried their past with them. Their thinking and mentality were slow to change. Since then, a long series of political conflicts has reduced the gap between the normative postulate of the Basic Law and the German reality. Germany today is the best in terms of democracy, the rule of law and civility that German history has to offer. This is something to be proud of. And no one can take that away from us.

Defending freedom – keeping the West together

Climate specialists talk of tipping points when global warming changes irreversibly from one stable state to another. A slow, barely perceptible increase in temperature then leads to a chaotic and erratic development. There are also tipping points in politics that trigger rapid changes. A situation that was hitherto regarded merely as a disruption to the old system suddenly takes on a new quality. There are indications that the West – the community of liberal democracies – is currently heading towards a tipping point of this type.

The American writer Anne Applebaum wrote in early 2016 that the West was possibly just three elections away from its end: the referendum on the United

Kingdom and the European Union, the US presidential election, and the French presidential election: Brexit, Trump, Le Pen – three coffin nails for the transatlantic alliance and the European Union. Two elections later, we already have one leg hanging over the precipice. In France, Emmanuel Macron rescued the EU. But the political developments in Hungary, Poland, Austria and Italy head in the opposite direction. Every future election is likely to be fateful. Will it be possible to stem the tide of extremism and shore up the democratic centre? Those who rely on countering the radicalization of the right with a radicalization of the left are playing with fire. We live in a period of far-reaching change. Politicians should not encourage the illusion that we can isolate ourselves from global economic competition, the digital revolution or worldwide migration movements. They must do everything to ensure that these phenomena are successfully managed. Europe is a continent with 500 million inhabitants, still the largest economic area in the world, with a great cultural tradition, high industrial potential and a wide-ranging research landscape. It would be a joke of history if we did not have the courage to enable future generations to live in freedom, security and prosperity.

This also applies to external challenges. While liberal democracies are full of self-doubt, the opponents of the West are making hay. Putin's dream is to isolate the United States from Europe and to extend Russia's sphere of influence to Western Europe. The new regime in the United States is a gift from heaven for the Kremlin. Donald Trump's destructive politics, his maverick approach and his evident aversion to multilateral commitments have encouraged those who have long called for the United States to distance itself more from Europe.

There is no reason why the EU should not assume greater responsibility for its own security policy. It is long overdue. But it would be an act of strategic stupidity to cancel the transatlantic alliance. America is bigger than Trump. NATO remains the cornerstone of European security, and the close economic ties with America are the driving force behind innovation and employment. Without the transatlantic alliance, Europe will become a minor player in world politics. In terms of our constitutional values, we still have more in common with North America than with any other world region. We must defend these common features as much as we can, in spite of and if necessary against Trump. The political line of separation does not run between Europe and America but between the defenders and opponents of an open society on both sides of the Atlantic. We need the transatlantic alliance of democrats more than ever.

In view of the uncertainty as to America's future path, Germany will play a key role in the future of the West. In Warsaw and Washington, Kyiv and Paris, politicians of the democratic centre, human rights defenders and intellectuals all see Germany today as a vital leading European power. It is a cause for concern that a right-wing populist party is also making trouble in Germany and gives rise to fear of an unstable Germany, wandering aimlessly between the West and Russia, rather than a strong and capable Germany.

Germany's long road west[1] is still embattled. Every German government is measured by its ability to protect liberal democracy, promote European integration and defend the transatlantic alliance, if necessary against the American president. Those who waver in this question are seen as a risk to our security and freedom.

There are two mistakes in critical situations that should be avoided if possible: panic and trivialization. We cannot avoid the fight for an open society. We will win it if we represent our values confidently and combine a love for freedom with a sense of justice.

What each one of us can do:

- Vote! Every vote counts. Suffrage is the basic law of democracy. Study the candidates and parties and make your choice. It is *not* immaterial who rules.
- Make a stand in everyday life! Speak up against injustice. Answer back when people are disparaged. Racist and misogynistic remarks are not trivial matters. Stand up for your convictions.
- You matter! Democracy depends on citizens who engage in public affairs. You can do this in many ways – through citizen initiatives, environmental associations, human rights organizations and political parties. Financial commitment also counts: support initiatives and projects through regular donations.
- Support independent critical journalism. In particular in times of uninhibited propaganda, half-truths and outright lies, there is a need for quality media that get to the bottom of things, conduct in-depth research and distinguish between information and opinion. Good journalism is not free. Quality newspapers and internet media have a price.
- Be sceptical of all miracle cures, simple explanations and political patent recipes. That is humbug. In a complex world, there are no simple solutions. Viable solutions must address conflicting aims and a variety of different interests.
- Defend the constitution! No one is above the Basic Law – no minister, no party, no religion, no company. An attack on the constitution is an attack on democracy.
- No violence! In a democratic constitutional state there is no justification for violence as a political instrument. Those who pursue their goals with violent means are destroying civilized political culture. Violence starts with speech, the debasement and dehumanization of opponents. Those who preach hatred sow the seeds of violence.

- Oppose the disdain for politics and parliaments. The disparaging talk of 'politicians' and 'parties' encourages anti-democratic sentiments. Criticism is the salt in the democratic soup; the sweeping disparagement of politicians and institutions poisons it.

Notes

Chapter 1 In Place of an Introduction: The Lie of the Land
1. See the standard work by the German-American historian Fritz Stern, *The Politics of Cultural Despair* (Berkeley, CA: University of California Press, 1961). Stern shows that Nazism did not suddenly appear but was the legacy of a long history of anti-liberal, *völkisch* and nationalist thinking.
2. 'Radical Chic: That Party at Lenny's' is the title of a famous essay by the American journalist Tom Wolfe in 1970 about the flirtation of liberal high society with radical ideas and movements. His prime example was a gala event to raise money for the Black Panther Party, a revolutionary movement in the United States, organized by Leonard Bernstein and his wife in their thirteen-room penthouse on Park Avenue in Manhattan. The guest of honour was John Cox, field marshal of the Black Panthers.
3. Hannah Lühmann, 'Warum haben linke Männer keine Eier?', *Welt*, 2016, https://www.welt.de/kultur/article 156325398/Warum-haben-linke-Maenner-keine-Eier.html.
4. Karl R. Popper, *The Open Society and Its Enemies, Vol. II – The High Tide of Prophecy: Hegel, Marx and the Aftermath* (London: Routledge, 1947), p. 71.
5. See Alexander Dugin, *The Fourth Political Theory*, trans. Mark Sleboda and Michael Millerman (London:

Arktos, 2012). Dugin's main enemy is the liberal universalism of the West, which he contests as a new form of totalitarianism. His writings are a chaotic mixture of Heidegger, fascist theoreticians such as René Guénon and Julius Evola, national Bolshevism and post-structuralism.

Chapter 2 Modern and Anti-Modern
1. Helmuth Plessner, *The Limits of Community: A Critique of Social Radicalism*, trans. Andrew Wallace (New York: Prometheus Book, 1999).
2. Quoted in Stern 1961, p. 259. In the early 1920s, Moeller van den Bruck also published the complete works of Dostoevsky in German, which were widely read and shaped the German image of Russia. Moeller described Dostoevsky as a kindred spirit in his dismissal of modern liberalism.
3. Popper, *The Open Society*, p. 60.
4. Dugin 2012, p. 329.
5. Arthur Moeller van den Bruck, *Das dritte Reich*, quoted in Stern 1961, p. 259.
6. Dugin, p. 259.
7. Ibid., p. 233.
8. Ibid., p. 310.
9. Dugin speaks in an earlier text of 'fascism – limitless and red'; see Andreas Umland, 'Faschismus à la Dugin', *Blätter für deutsche und internationale Politik* (December 2007, pp. 1432–5); https://www.blaetter.de/archiv/jahrgaenge/2007/dezember/faschismus-a-la-dugin.

Chapter 3 The Long View of Democracy
1. Anthony Giddens, *Beyond Left and Right – The Future of Radical Politics* (Cambridge: Polity Press, 1994).
2. Carl Schmitt, *The Crisis of Parliamentary Democracy*, trans. Ellen Kennedy (Cambridge, MA and London: MIT Press, 2000), p. 10.
3. Ibid., p. 15.
4. Popper, *The Open Society*, p. 37.
5. A scene from Schiller's drama *Demetrius*. The prince opposes his fellow nobles in the Polish imperial assembly,

who are in favour of entering into a war with the Russian Empire.
6 Schmitt, p. 5.
7 Ibid., p. 4.
8 Hannah Arendt's message is 'The meaning of politics is freedom'; see Hannah Arendt, *Was ist Politik? Fragmente aus dem Nachlass*, ed. Ursula Ludz (Munich/Zurich: Piper Verlag 1993), p. 28.

Chapter 4 The Left and Democracy
1 Theodor W. Adorno, *Minima Moralia: Reflections from a Damaged Life*, trans. E. F. N. Jephcott (London and New York: Verso, 2005), p. 39.

Chapter 5 Rise of the Anti-Liberals
1 Thomas Assheuer, 'Etwas Besseres als die Freiheit finden wir überall: Warum das autoritäre Weltbild rechtspopulistischer Parteien so erfolgreich ist', *ZEIT* 7/2016.
2 See interview with the 'AfD philosopher' Marc Jongen in *ZEIT*, http://www.zeit.de/2016/23/marc-jongen-afd-karlsruhe-philosophie-asylpolitik.
3 Quoted in James Pethokoukis, 'Is it no longer "the economy, stupid"?', American Enterprise Institute, 27 July 2016.

Chapter 6 The Migration Battlefield
1 'Letzte Hoffnung Europa', 3sat website, 16 October 2013; see http://www.3sat.de/page/?source=/nano/gesellschaft/172745/index.html.

Chapter 7 Dealing with Islam
1 All quotes from the German Wikipedia entry on *Charlie Hebdo*; see https://de.wikipedia.org/wiki/Charlie_Hebdo.

Chapter 8 No Empathy for Freedom: The Germans and Ukraine
1 See Timothy Snyder, *Bloodlands: Europe between Hitler and Stalin* (London: Vintage, 2011). Snyder's pioneering study is essential reading for anyone wishing to know

more about the shadow cast on the present day by the history of Central Europe between 1933 and 1945. The figures quoted here can be found on p. 404.

Chapter 9 The Russian Complex
1. Details can be found in the fascinating study by the American political scientist Karen Dawisha, *Putin's Kleptocracy: Who Owns Russia?* (New York: Simon and Schuster, 2015).
2. 'Putin-Freund Sergej Roldugin: Melodien für Milliarden', *Sueddeutsche.de*, 10 April 2016, http://www.sueddeutsche.de/politik/panama-papers-putin-freund-sergej-roldugin-melodien-fuer-milliarden-12943661.
3. Marx wrote a whole series of articles in 1853 for the *New York Daily Tribune* on the Russian occupation of the principalities of Moldova and Walachia, which were under Turkish protection. In them, he criticizes the appeasement policy of Britain and France, stating that Russia's expansion policy is based on the cowardice of Europe; see Karl Marx and Friedrich Engels, 'Russian Policy against Turkey – Chartism' in *Karl Marx Frederick Engels: Collected Works, Vol. 12 1853–1854* (London: Lawrence & Wishart, 1979), pp. 164–73, 167.

Chapter 10 Modernity and Its Discontents
1. Goethe uses the term 'veloziferisch', a combination of the Latin *velocitas*, speed, and Lucifer, the traditional Christian name for the Devil.

Chapter 12 Civilizing Capitalism
1. A collective term for a wide range of investments following strict ecological and social criteria to a greater or less extent. The spectrum ranges from the exclusion of certain sectors, such as the armaments industry, to demanding sustainability criteria; see https://www.nachhaltigkeit.info/artikel/marktentwicklung_und_marktsegmente_1666.htm.
2. Popper, *The Open Society*, p. 117.
3. Ibid., p. 224.

Chapter 15 What is at Stake
1 The title of Heinrich August Winkler's two-volume work on German history from 1806 to 1990.